ASK A SCIENCE TEACHER

NEW YORK

ASK A SCIENCE TEACHER

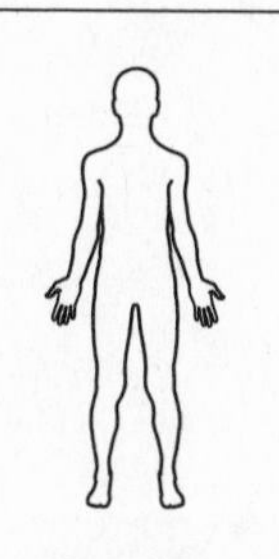

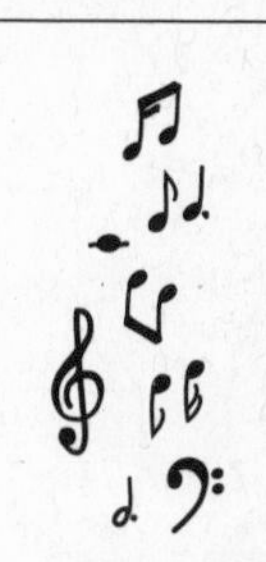

250 Answers to Questions You've Always Had About How Everyday Stuff *Really* Works

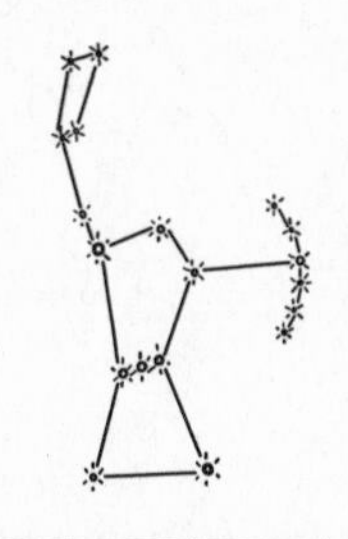

Larry Scheckel

Ask a Science Teacher: *250 Answers to Questions You've Always Had About How Everyday Stuff Really Works*

This is an extensively revised and updated edition of *Ask Your Science Teacher*, originally published in 2011 by CreateSpace Independent Publishing Platform.
Illustration on page 235 based on an original design by Colin Bradley of PyroUniverse.com
Illustration on page 256 courtesy of Wikimedia Commons User Cepheus
Illustration of Gun-Type Fission Weapon on page 272 courtesy of Dake, FastFission, and Howard Morland used under Creative Commons Attribution 2.0 Generic license
Illustration of Implosion Nuclear Weapon on page 273 courtesy of Ausis and FastFission used under Creative Commons Attribution-Share Alike 2.5 Generic license
Illustrations on pages 1, 75, 105, 147, 197, 231, 259, 285, 297, 323, and 333 courtesy of Karen Giangreco.

The Experiment, LLC
220 East 23rd Street, Suite 301
New York, NY 10010-4674
www.theexperimentpublishing.com

Library of Congress Cataloging-in-Publication Data

Scheckel, Larry, author.
Ask a science teacher : 250 answers to questions you've always had about how everyday stuff really works / Larry Scheckel.
pages cm
ISBN 978-1-61519-087-4 (pbk.) -- ISBN 978-1-61519-179-6 (ebook)
1. Science--Miscellanea. I. Title.
Q173.S288 2013
500--dc23
2013022014

ISBN 978-1-61519-087-4
Ebook ISBN 978-1-61519-179-6

Cover design by Howard Grossman
Text design by Pauline Neuwirth, Neuwirth & Associates, Inc.

Manufactured in the United States of America
Distributed by Workman Publishing Company, Inc.
Distributed simultaneously in Canada by Thomas Allen and Son Ltd.
First printing November 2013
10 9 8 7 6

This book is dedicated to my wife, Ann, for her suggestions, constant encouragement, joy, and inspiration.

And to my deceased parents, Alvin and Martha Scheckel, who made me what I am today.

Contents

Chapter 4: Technology 147

Preface

Why study science? That's a question that many students, and indeed adults, ask. It is a very good question, because it gets at the heart of what science is: think of science as a tool for answering the basic questions of how things work and why things are the way they are. The true value of science is learning about the world around you; a fundamental understanding of the principles of science allows us to delve deeper into the mysteries of the universe.

Different branches of science explore different realms of our world. Physics is the study of matter and energy and how they interact with each other. Newton's laws of motion and gravitation, for example, are central to grasping the mechanisms of rockets, satellites, roller coasters, and cars. Physics can allow a person to appreciate friction when applying disc brakes on a car or when trying to minimize this speed-reducing force while constructing a pinewood derby race car.

Chemistry is the study of the composition, properties, and behavior of matter. Everything we see, feel, smell, taste, and touch involves chemistry and chemicals. Consider a choking green gas such as chlorine. Ponder sodium, a nasty metal that is silvery and soft and that reacts violently with water. But put them together and we get a substance that no cook could live without—salt.

Biology is the study of living things—their structure, function, growth, classification, and reproduction. Here lies a vast study of how cells divide, why we have blood types, how DNA holds the secrets of genetic coding, how tissues and organs age, and how the brain of a dog's owner is different from the brain of the dog.

We've just mentioned the Big Three sciences. Within these main categories is an endless world of in-depth study in over fifty specific branches of science, from astronomy to geology to zoology.

In addition to science being just plain fun, there are some important and practical reasons for studying science, for the benefit of everyone, not just the few or the elite. Whether a singer, janitor, farmer, or nuclear physicist, it is important for a person to think scientifically and make decisions based on solid information about the world. Even arcane questions about whether to smoke or not, what to eat, or which car to buy depend on scientific facts. Some people, like policy makers in business and government, make decisions that affect many other people. They deal with national and international questions of population growth, environmental concerns, nuclear power, climate change, and space exploration and with state and local questions of water and sewer treatment, highway and bridge construction, what trees to plant along the boulevards, and school construction, to name a few, all of which are best resolved if the solutions are based on scientific principles. And responsible voters need enough science sense to be able to evaluate candidates' positions on these questions.

But science isn't merely utilitarian; in my experience as a science teacher, learning about science is also great fun. Jules-Henri Poincaré (1854–1912), the French engineer, physicist, and mathematician, said, "The scientist does not study nature because it is useful; he studies it because he delights in it, and he delights in it because it is beautiful. If nature were not beautiful, it would not be worth knowing, and if nature were not worth knowing, life would not be worth living." It is this beauty and delight in nature, and in the science behind it, that inspired me to start writing a science column, called "Ask Your Science Teacher," in 1993. I live in the small city of Tomah, Wisconsin, population 8,000. Our newspaper, the *Tomah Journal*, is published twice a week, with my column appearing every Thursday. In a way, I've become my town's resident science expert—I'm their science teacher.

Before I wrote a single column, I had to grapple with one very important issue: Where to get some questions to answer? I decided to ask other teachers in the area to pass out sheets of scrap paper to their students and have them write down any questions they had about things that they were curious about, or that had been bothering them, or that they always wondered about. Within two weeks I had about 130 questions to choose from. I selected ten and started to formulate answers, which would become my very first columns.

Children often make the best questioners, because they have open minds and fresh eyes. Adults often take more for granted, so they tend not to question the fundamental, everyday stuff we encounter all the time but never think to ask about. Children, on the other hand, are curious about everything—their bodies, the solar system, and the things they are learning about in school—and they aren't afraid to ask.

This book, *Ask a Science Teacher*, is for everyone who has not had the chance to be in my classroom or read my column. It has been twenty years now since my first column, and I've since written more than 550; I selected 250 of my best to include here.

There are many resources that I have found helpful in ensuring that I'm crafting completely reliable answers, especially for questions concerning the biological, chemical, and Earth sciences, which fall outside the realm of my expertise, the physical sciences. In some cases, I turned to magazines, such as *The Science Teacher*, *Science and Children*, *Science Scope*, and *The Physics Teacher*. I also found the Internet to be a valuable tool. And, at times, I contacted doctors, engineers, lawyers, business people, and manufacturing plant managers for help and suggestions.

When doing a weekly newspaper column, one encounters some surprises. I found I got some questions that can't be easily answered or perhaps don't even have answers. How would I even begin to answer questions like: Who made God and why? Why are people mean to each other? Is there a chance that gravity may fail? Why do pigs snort? What is the meaning of life? Why can't cows talk? (We do have a lot of Holsteins in Wisconsin!) or Why can't chickens and turkeys fly?

I sincerely hope this book provides a "good read" and that it engages the mind and piques curiosity. It is my wish that you, like Jules-Henri Poincaré, will delight in science because it is beautiful and worth knowing more about.

1

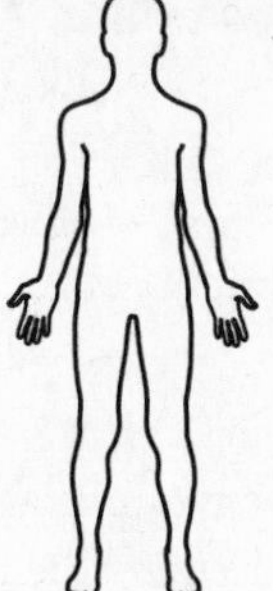

The Magnificent Human Body

1. How many cells are in your body?

There is no real consensus on the number of cells in the human body. Estimates put the number between ten trillion and one hundred trillion. A trillion is a million million—it's a word that crops up when we talk about the size of our national debt! The number of cells depends on the size of the person: bigger person, more cells. Also, the number of cells in our body keeps changing as old cells die and new ones form.

Cells are so small that most can only be seen through a microscope. Every cell is made from an already existing cell. Each cell in the body behaves like a little factory and has two major components, the cytoplasm and the nucleus. The cytoplasm contains the structures that consume and transform energy and perform many of the cell's specialized functions, including storing and transporting cellular materials, breaking down waste, and producing and processing proteins. The nucleus is the control center and contains the genetic information that allows cells to reproduce. The mitochondrion (plural *mitochondria*) in the cell is the factory where food and oxygen combine to make energy. Human cells and other animal cells have a membrane that holds the contents together. This membrane is thin, allowing nutrients to pass in and waste products to pass out. Food is the energy the cell needs. Each cell needs oxygen to burn (metabolize) the nutrients released from food.

The body has some cells that do not experience cell division. And red blood cells and outer skin cells have cytoplasm but do not have a nucleus.

In the cell, the process is called respiration. Oxygen breaks down the food into small pieces. The oxidizing of the food molecules is turned into carbon dioxide and water. Water makes up about two-thirds of the weight of the cell. The energy released is used for all the activities of the cell. The cell membrane has receptors that allow the cell to identify surrounding cells. Different kinds of cells release different chemicals, each of which causes certain other types of nearby cells to react in certain ways. Within each of these different cells are found twenty different types of organelles, or structures.

Slightly over two hundred different kinds of cells make up the human body. The shape and size of each type of cell is determined by its function. Muscle cells come in many different forms and have many different functions. Blood cells are unattached and move freely through the bloodstream. Skin cells divide and reproduce quickly. Some cells in the pancreas produce insulin, others produce pancreatic juice for digestion. Mucus is produced in cells in the lining of the lung. Our lungs also contain alveolar cells that are responsible for taking in gas from the blood. The cells that line the intestine have extended cell membranes to increase the surface area, helping them absorb more food. Cells in the heart have a large number of mitochondria to help them process a lot of energy, because they have to work very hard.

Nerve cells generate and conduct electrical impulses; for the most part, they do not divide. Each nerve cell has a specific place in our nervous system. Nerve cells outside of the brain are very long and have the task of passing signals between the brain and the rest of the body, allowing us to move our muscles and sense the world around us. The rest of our nerve cells—about one hundred billion of our body's cells—are brain cells.

Brain cells are the most important cells in our bodies. It is our brain that defines who we are. Brain cells in children under five do have the ability to reproduce, to some extent. However, we are naturally losing brain cells all the time. The best estimate of normal brain cell loss is put at nine thousand per day. That may seem like a large number, but remember that the brain has 100 billion cells, so a nine-thousand-cell loss per day is not that great. Inhalants, such as glue, gasoline, and paint thinner, cause brain cell loss at thirty times the normal rate. Excessive alcohol use is a big contributor to brain cell damage.

Cells that all do the same job make up tissue, such as bone, skin, or muscle. Groups of different types of cells make up the organs of the body. Different organs grouped together form a system, such as the digestive system or the circulatory system. All the systems working together make up a healthy human body.

Cells live, of course, but cells also die. Liver cells last about a year and half. Red blood cells live for 120 days. Skin cells are good for 30 days. White blood cells survive for thirteen days. And it turns out that the great majority of cells in the human body are bacterial cells, and most are beneficial. It is hard to believe that the average adult loses close to 100 million cells every minute. The good news is that the body, through cell division, is replacing those lost 100 million cells every minute. And in any case, even 100 million cells is only a small fraction of the trillions of cells that make up our bodies.

2. Why do the young and the elderly get sick more easily?

Babies get sick more often than older children or adults because their immune systems are not fully developed and functioning at full capacity. The common cold, which is an infection of the respiratory system caused by a virus, is the most frequent malady. Doctors say that normal, healthy babies get up to about seven colds before they reach their first birthday. Another common affliction is the flu, caused by a different family of viruses, which bring on high fever, chills, fatigue, and sometimes digestive symptoms like vomiting and diarrhea, in addition to the respiratory symptoms of a cold.

Another reason babies get sick so often is that they are frequently around other children, often siblings, and this exposes them to viruses and bacteria in school and daycare. Children in schools and daycare get more colds, runny noses, and ear infections than children cared for at home. However, their earlier exposure to these diseases also leads them to develop immunity earlier.

Babies are also curious about the wide, wonderful world they are born into. So they will stick anything and everything into their mouth as a means of exploring that world. You can imagine the enormous amount of germs that ride along.

Furthermore, babies have not developed the immunity to the many different viruses that cause colds, because they haven't had time to acquire the antibodies to fight off viruses. Babies do have some of their mothers' antibodies when they are born, which were transmitted through the placenta during pregnancy. This kind of immunity isn't permanent, but breastfeeding can extend it, because many of the mother's antibodies are present in her milk. This is why breast-fed babies tend to have fewer colds and flu symptoms than bottle-fed babies. Babies, like other people, also develop their own antibodies in response to germs they are exposed to; in fact, it's a mistake to try to eliminate all pathogens from a baby's environment.

Winter is the toughest time for babies, because it is the season when colds spread nationwide. Also, in winter people spend more time indoors, where viruses are more likely to spread from one person to another. The less humid air of indoor heating dries nasal passages, which allows viruses to thrive.

All people, both adults and children, are susceptible to bacterial and viral infections. Bacterial infections include meningitis, cholera, bubonic plague, tuberculosis, diphtheria, and anthrax. Vaccines for these dreaded ailments were developed decades ago. But when the very young and very old get sick, it is most often from viruses. A prime example is the common cold.

There is no cure for the common cold, because many different viruses cause colds and even if a medicine is developed for one of them, people would still catch colds from other viruses. Many people who have colds, or whose children have colds, ask their doctors for antibiotics, because they don't understand that these drugs don't work against viruses. But there are medicines that can relieve the symptoms of colds and flu so babies can get better sooner and not suffer as much. Recent research has developed medicines against some viruses; for example, the vaccine that helps prevent the flu can also treat it if given soon enough after a person develops symptoms.

Elderly people get sick more often because their immune systems are weakened or breaking down. They also tend to have existing conditions that make them more vulnerable. Some have heart disease, kidney problems, asthma, diabetes, and a whole host of illnesses that no one looks forward to. Many of these diseases, as well as their treatments, suppress the immune system.

That's why the Centers for Disease Control and Prevention (CDC) advises that children under five years of age and people over sixty-five years of age have flu shots when each new strain begins to spread. Most people who contracted the H1N1 swine flu virus in 2009 came down with a mild illness, but the fatality rate was high. According to the World Health Organization (WHO), 284,500 died from the H1N1 virus, the majority being from Africa and Southeast Asia.

3. What are birthmarks?

A birthmark is a colored spot on or just under the skin. Most birthmarks show up when a baby is born. Some are noticed shortly after the baby is born. Some birthmarks fade away as the child grows up, but some stay and get bigger, thicker, and darker.

Nearly all birthmarks are harmless and painless. They can be almost any size, shape, or color.

Birthmarks have two primary causes: blood vessels that bunch together or do not grow normally and extra pigment-producing cells, or melanocytes, in the skin. Doctors don't know what's responsible for these two causes, but many think there is a genetic component involved.

The most common birthmark is the port-wine stain. The stain is usually pink-red at birth and tends to become red or purple as a person ages. Port-wine stains, caused by blood vessels that do not grow normally, can have various sizes and shapes. Port-wine stains most often show up on the face, back, or chest. The strawberry birthmark is another that is found on newborns. It is also caused by a clumping of blood vessels that do not grow normally. Mongolian spots are benign congenital birthmarks found mostly on East Asians. Originating on the lower back, these bluish spots disappear by the time the child reaches age five. A salmon patch is a very common birthmark, occurring on 75 percent of newborns. It is caused by dilation of tiny blood vessels. Most salmon patches disappear by age one or two. Stork marks appear on the back of the neck, middle of the forehead, or upper eyelids. They vanish by the time the child is two years old.

The downside of birthmarks is that kids have to live with the teasing, ribbing, and cruel remarks of classmates. Some kids can go through a miserable childhood enduring the slings and arrows of their peers. But there is some good news. Makeup creams can hide many birthmarks on the face and neck or make them less noticeable. Others can be removed by surgery or lightened with a laser, but these treatments can be painful. Since most birthmarks are harmless, most are not treated.

4. Why is blood red?

Blood is red because hemoglobin, a protein in red blood cells that binds oxygen and carbon dioxide, contains chemical compounds called hemes, and a heme is a blood pigment that contains iron, which is reddish in color. There are about thirty-five trillion red blood cells—tiny, round, flat disks—circulating in our blood at any one time—that's thirty-five followed by twelve zeroes. And each red blood cell typically has more than 250 million hemoglobin molecules, each with four heme groups!

Blood is pumped by the heart and circulated around the body through blood vessels. Blood is bright red when the hemoglobin picks up oxygen in the lungs. The red blood cells carry the oxygen, bound to their hemoglobin, to the rest of the body through arteries and capillaries. Carbon dioxide from the body's cells returns to the heart through capillaries and veins. The darker venous blood carries the carbon dioxide from the tissues to the lungs, which expel them.

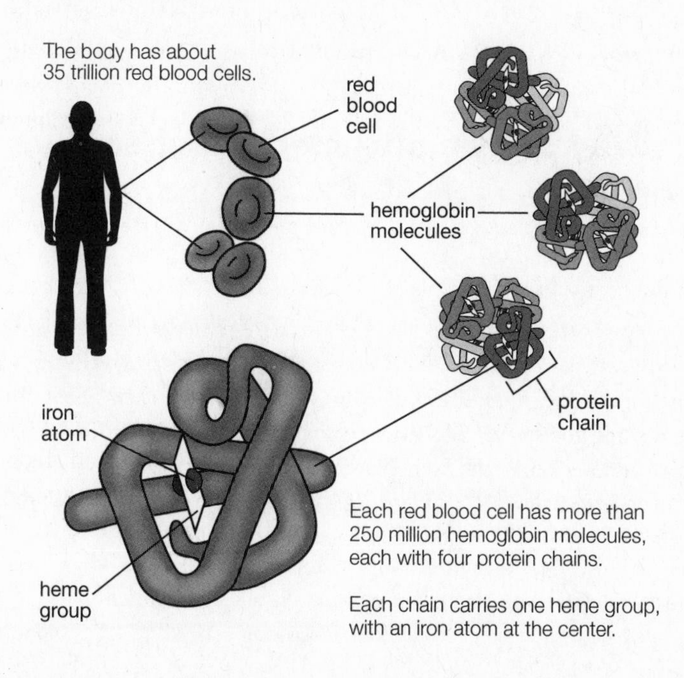

The blood coursing through our body's plumbing of arteries, veins, and capillaries contains many different materials and cells. Plasma, the liquid part of blood, is a light yellow color, denser than water, and carries proteins, antibodies to fight diseases, and fibrinogen, which helps the blood clot. Plasma also has carbohydrates, fats, and salts. Young red blood cells mature in the marrow of the bone. Red blood cells have a life expectancy of about four months. Then they are broken up in the spleen and replaced by new blood cells. New cells are constantly replacing old cells. Our blood also contains several types of white blood cells. When a germ infects the body, some white blood cells race to the scene and produce protective antibodies that overpower the germs, while other white blood cells surround and devour them.

The average adult has between eight and twelve pints, or four to six quarts of blood. If a person loses a significant portion of their blood supply, they go into shock and die. This can be prevented by transfusing blood from another person with a matching blood type (see page 13). The first blood transfusion on record took place in 1665. Richard Lower of Oxford, England, took blood from one dog and put it in another dog. The first known human-to-human blood transfusion took place in 1795 in Philadelphia.

5. Why are we attracted to unhealthy foods?

Some short answers: lots of junk foods have loads of sugar. Many junk food items are brightly colored, which attracts our (especially children's) attention. People like finger foods, such as burgers, hot dogs, and fries. Advertisers target children, making junk food more attractive than is healthy from a young age. Deep-fried foods are tastier than bland foods, and children and adults develop a taste for such foods. Fatty foods cause the brain to release oxytocin, a powerful hormone with a calming, antistress, and relaxing influence, said to be the opposite of adrenaline, into the blood stream; hence the term "comfort foods."

We may even be genetically programmed to eat too much. For thousands of years, food was very scarce. Food, along with salt, carbs, and fat, was hard to get, and the more you got, the better. All of these things are necessary nutrients in the human diet, and when their availability was limited, you could never get too much. People also had to hunt down animals or gather plants for their food, and that took a lot of calories. It's different these days. We have food at every turn—lots of those fast-food places and grocery stores with carry-out food.

But that ingrained "caveman mentality" says that we can't ever get too much to eat. So craving for "unhealthy" food may actually be our body's attempt to stay healthy.

Food manufacturers put color additives in their foods. Bright, vibrant, saturated colors look more appealing to consumers. A bright red apple is more appealing than a dull red or green apple. A key to survival in olden times was the ability to recognize foods that contained usable energy or nutrition. People needed to be able to recognize foods that contained many calories, would support healthy brain function, harbored healing medicines, and boosted the immune system. Many of those natural foods often appeared in bright colors, such as apples, oranges, bananas, carrots, and berries. Color was a reliable indicator of a healthful food. Indeed, when apples, bananas, or berries spoil, they lose their bright colors. So food-manufacturing companies are exploiting what was once a well-based notion that colorful foods are healthy foods!

Additives make foods taste better, look better, and last longer on the shelf. Experts agree that all those additives can't be good for us. Some additives come from coal tar and petrochemicals. Our bodies are not made to ingest crude oil. Some additives have been shown to be safe, but many have not even been tested. There is growing suspicion that all those additives are responsible in part for a rise in child obesity, attention deficit hyperactivity disorder (ADHD), and questionable behavior patterns. Some food dyes, such as Blue #2, Yellow #5, and Red #40, have been linked to cancer and ADHD.

At the same time, some foods have their nutrients subtracted. White flour starts out as whole wheat flour, but the manufacturers strip out its fiber, along with many nutrients. Then they "enrich" it by adding back some nutrients, but it's not the same as the original whole grain.

Another side effect of our taste buds' being attracted to unhealthy foods is the huge increase in type 2 diabetes, from a double whammy of eating too much and eating the wrong kinds of food. This is a type of diabetes that used to appear after adulthood—in fact, it's often called "adult-onset diabetes." But lately, more and more cases begin during childhood and adolescence, and this has been linked to the increase in childhood obesity. Anything made with a high sugar content and/or white flour spells trouble for the diabetic. That means cookies, cake, candy, soda pop, ice cream, and pastries. These foods are loaded with fat and salt that contribute to high blood pressure and heart disease. Most of us like to eat these delicious foods. So experts say it goes back to Aristotle's advice for a happy life: the Golden Mean—moderation in everything.

6. Why do we have nightmares?

The definition of a nightmare is a really bad and distressing dream that causes a strong feeling of fear. A nightmare quickens a person's pulse and makes them sweat. Sometimes, the sleeper feels so frightened and threatened that they wake up. Sleep experts estimate that 30 to 50 percent of all kids have some nightmares, but luckily, they usually grow out of them. The most common nightmare I had as a kid was that of being chased, but I can't recall why I was being pursued.

Nightmares can be fairly long and complex. The person senses a threat to safety or life. As the threat increases, so does the sense of fear. A person tends to wake up just as the threat or danger reaches its peak.

About 3 percent of young adults have frequent nightmares. One in two adults have a nightmare on occasion. An estimated 2 to 8 percent of adults are plagued with frequent nightmares. Stress, depression, and anxiety are commonly associated with nightmares in adults. A major life-changing event can cause them, such as loss of a job, financial worries, marital difficulties, death of a spouse, or moving to another house. Alcohol abuse or abrupt withdrawal from alcohol can also lead to nightmares.

Nightmares occur during the rapid eye movement (REM) phase of sleep. REM is about two hours of a normal night's sleep, but other phases of sleep break that time up into four or five episodes. REM phases become progressively longer as the night progresses, and so you may find you experience nightmares most often in the early-morning hours of REM sleep. One can have four to five episodes of REM but still usually have nightmares in the later stages of sleep

Traumatic experiences, such as surgery, brain injury, war, and combat, with their attendant post-traumatic stress disorder (PTSD), can bring on nightmares. Stress is thought to be the most common source of nightmares, so relaxation techniques, such as yoga and meditation, have proven helpful. And eating right before going to bed can increase the frequency of nightmares, since eating increases metabolism and brain activity.

There is no diagnostic test for nightmares. Persistent disorders surface when people report them to a family doctor or a psychiatrist.

7. How do braces make your teeth straight?

Braces have two results: They make your teeth straight, and they make your parents get a second job to pay for them!

Actually, straight teeth help a person effectively bite, chew, and speak. Teeth that are properly aligned tend to look better and work better. Straight teeth can also prevent decay by giving plaque fewer places to hide. That nasty plaque can lead to gingivitis or more serious periodontal (gum) disease. Protruding front teeth have a good chance of being broken or fractured in an accident. Crooked teeth can cause abnormal wear on tooth surfaces, misalignment of jaw joints, neck and facial pain, and even headaches.

An attractive smile is a positive side effect of having straight teeth, bolstering self-esteem and self-confidence. Good teeth can be a social and career booster.

So how do braces work? They put pressure against the teeth, which makes them gradually move over a period of time. The pressure comes from an "arch wire" attached to the braces that runs on the outside of the teeth. The top teeth and the bottom teeth form separate arches. The arch wire is springy, and when attached to the braces on the teeth, it becomes deformed or bent along the path of the uneven teeth, and so it exerts a gentle force on the teeth to gradually move them to their desired position. Sometimes the braces are put on the inside of the teeth, where they are invisible. But that can cause problems, such as speech impediments and irritation of the tongue.

Older braces had the arch wire connected to large metal bands cemented around each tooth. Today's braces have the arch wire attached to tiny brackets that are cemented to the front of the teeth, with a slot on the outside to hold the wire. In both kinds, as the teeth get closer to the desired positions, the arch wire needs adjustment from time to time to apply continuous pressure to move them over the next interval.

Newer arch wires are made from Nitinol (**ni**ckel **ti**tanium **n**aval **o**rdnance **l**ab) wire, a space-age metal alloy that NASA uses to deploy satellite antennas that are folded up during launch.

The Naval Ordnance Lab, located in the Maryland countryside, established in the early 1920s, carried out important work in countering Germany's magnetically activated mines dropped by aircraft into Allied shipping lanes in WWII. After WWII, the Naval Ordnance Lab did basic research on metals, explosives, wind tunnels, underwater weapons, and radiation detection.

At room temperature, the Nitinol is very flexible and will hold a deformed configuration, but when heated, the nickel titanium (Ni-Ti) wires return to their original shape. So, in the heat of a person's mouth, the alloy wire that has been bent along the uneven teeth continually applies pressure to the teeth as it returns to its former contour.

After applying an etching material to each tooth, which prepares the enamel for a bonding agent that will hold each bracket to the tooth, the orthodontist rinses the etch and applies a sealant chemical, then applies the brackets to the tooth surfaces. The bonding material does not harden, or set, until the orthodontist shines ultraviolet light on it; this allows time for any last-minute adjustments on the placement of the bracket before attachment of the arch wire to the brackets.

The first arch wire is round, thin, and flexible. In subsequent visits, the orthodontist uses rectangular, larger-diameter wires. The wire slot of the bracket is rectangular, so a rectangular wire fits more snugly into the bracket and exerts a greater force. Braces are generally left on for a tad less than two years. To keep the teeth straight after removal of the braces, people wear a retainer. The retainer holds the teeth in their correct positions while bone fills in around them. A retainer is worn for three to six months. The orthodontist may recommend wearing the retainer only at night, especially during those later months.

About 4.5 million people in the United States wear braces. Most are teens, but 20 percent are older adults. Some couldn't afford braces when they were younger; some decided they wanted straighter teeth or a better smile, or had a dentist who recommended braces.

8. Why does the human body reject all other blood types except type O?

That is not exactly the case; it depends on which blood type the recipient has. A blood type, or group, is a classification of blood based on whether or not inherited antigens reside on the surface of red blood cells. Antigens are foreign substances that induce an immune response and interact with specific antibodies in our immune systems.

In 1900, Austrian doctor Karl Landsteiner found a basis for classifying human blood into four groups. He discovered that red blood cells could carry two different antigens, which he termed A and B. Each of us has two ABO blood-type alleles (different forms of the same gene), because we each inherit one blood-type allele from our biological mother and one from our biological father. The presence or absence of the A and B antigens makes for four possible blood types: A, B, AB, and O. A person with type A blood has red blood cells that carry only antigen A. One with type B blood has red blood cells that carry only antigen B. An individual with AB blood has both antigens, and one with type O has neither. There is another antigen on red blood cells, called the Rh factor. This antigen is named Rh for the Rhesus monkeys the experiments that discovered it were done on. People who have the Rh antigen are said to be Rh positive. Those who have the Rh antigen missing are called Rh negative. Blood banks group people on the basis of their ABO and Rh factors into categories like AB negative or O positive.

A test called an antibody screen, performed on patients who may require red blood cell transfusions, detects most red blood cell antibodies.

When a person receives a transfusion of blood, their body recognizes the received blood as either "foreign" or "familiar." A person whose blood type is O will reject all other blood types, because their body does not recognize the A and B antigens. This sets off an immune response in which antibodies attack the new blood, either destroying its red cells or causing it to clot, which, unfortunately, often leads to the death of the individual. However, when a person with the A or B blood type needs blood, they can receive their own blood type or type O, and a person with type AB blood, known as the universal recipient, can receive any of the four types.

Blood type varies by ethnicity. Type O blood is carried by 45 percent of the population, 35 percent have Type A, 15 percent have Type B, and 5 percent carry Type AB. Type O is called the universal donor, because it is the only blood type that can be transfused to patients with all other blood types in the ABO group. So blood banks like to have a lot of type O donors. About half of all the blood ordered by hospitals is type O. Rh-negative blood is like the type O of the Rh blood group system: people with either Rh-positive or Rh-negative can safely receive it, while it is not safe for people with Rh-negative blood to have Rh-positive blood transfused.

9. Why do we cry?

Oh, I know what you mean. I just read my 401(k) statement! Actually, tears are good. Tears clean and lubricate the eyes. While most animals have a system to keep their eyes moist, we humans are the only mammals that cry emotional tears, which are triggered by a part of the brain that's responsible for feelings of sadness. This melancholy then sends signals to the endocrine system to release hormones around our eyes that cause tears.

Strong emotions can come from being very happy or very sad, or from enduring pain or being under a lot of stress. These emotions can result from getting an F on your report card or receiving your tax bill in the mail, or from lost love, a tear-jerker movie, or overwhelming joy. In addition to emotional tears, there are tears caused by irritation. If we get something in our eyes, such as a piece of dust or debris, producing tears helps wash it out of the eye.

Vision is one of God's greatest gifts to us. So it is in our interest to have a basic understanding of vision, how the eye works, and how to take care of these wonderful gifts.

The tear glands, or lacrimal (also spelled lachrymal) glands, are located under the upper eyelid. Tears flow through tiny ducts that secrete tears onto the eyeball in a process called lacrimation (or lachrymation). The tears form a film that coats the eye with three distinct layers. The innermost layer, which covers the eyeball, is made of a protein called mucin. This layer, secreted by the conjunctiva, coats the cornea and permits an even distribution of the next layer, which is the aqueous layer, made of water and proteins. The outermost layer contains oils that prevent evaporation of the other layers. Every time a person blinks, the tear film is spread over the eye to keep it moist and free of dust and other irritants.

Afterward, tears drain into two tiny openings on the edges of the upper and lower eyelids at the inner corner of the eye. From there, they are channeled by the tear ducts into the nasal cavity, which drains into the throat, and are swallowed. If there are too many tears, they overflow the lower lid and run down the cheek.

Our vision gets blurry when we cry because light has to get through all those layers to get to the retina of the eye. Our tears distort the light we see.

Tears have the job of protecting our eyes, and, surprisingly, they can work in quite complex ways. In addition to keeping our eyes from drying out, tears have a bit of salt in them, which helps prevent infection. They even carry oxygen and some nutrients to the eye's surface. If something gets in one of our eyes and irritates it, the tear glands flood our eyes with tears and try to flush away the invader. And tear production goes on even when we are sleeping.

People can have trouble with their tear plumbing system. Dry eye syndrome is a common condition, in which the eye is not being kept wet enough to be comfortable. Using artificial tears is the most common treatment. The openings (puncta) that conduct tears out of the eyes can become blocked. A clean, warm, wet washcloth placed over the closed eyes is the recommended treatment. The warmth may open clogged oil ducts. And there is a whole host of other maladies associated with the tear system. Doctors can diagnose and recommend treatment.

Health professionals say that crying is beneficial to health and mental well-being. A good cry is great for body and mind.

10. What is juvenile diabetes, and how can you get it?

Type 1 diabetes, which used to be called juvenile diabetes, is a chronic, lifelong autoimmune disease in which the beta cells in the pancreas's islets of Langerhans produce too little insulin to regulate blood-sugar levels. Without adequate insulin, glucose (sugar) builds up in the bloodstream instead of going into the body's cells. The body is not able to use this glucose for energy despite the fact that there are high levels in the bloodstream, which leaves the person feeling hungry. Additional symptoms include thirst, increased urination, weight loss, nausea, and fatigue.

There is one new case per year for every seven thousand children, but the exact cause of juvenile diabetes is a mystery. Genetics play a role, but cold weather and viral infections are thought to be possible environmental factors as well and may trigger an autoimmune phenomenon leading to the destruction of beta cells in the pancreas.

Seven percent of the population of the United States, or twenty-one million people, have diabetes. About 5 to 10 percent of those cases are type 1 diabetes. Most Americans who are diagnosed with diabetes have the less serious type 2, formerly called adult-onset, diabetes. The good news about type 2 diabetes is that many cases can be controlled by diet, weight management, and exercise. But patients often still need oral medication or injected insulin to achieve effective control of their condition. A recent phenomenon has been the marked increase in the incidence of young people being diagnosed with type 2 diabetes (previously rare), which correlates with the "epidemic" of childhood obesity. Many of them are within the age range once associated with type 1 diabetes, and this helped lead to the dropping of the juvenile- and adult-onset labels.

A diagnosis of diabetes is confirmed by a fasting blood test—a patient does not eat anything for twelve hours before the test. Water, tea, or coffee is acceptable, but skip the milk and sugar. Treatment takes two forms: insulin injection to regulate blood sugar levels and treatment for diabetic ketoacidosis (DKA), which can occur when blood-sugar levels do get too high (hyperglycemia). At around 240 milligrams per deciliter, the body looks for other energy sources and uses fat as the fuel source.

The fats are broken down and acids called ketones build up in the blood and urine. DKA is a dangerous situation and accounts for most of the hospitalizations due to diabetes. DKA can lead to heart and kidney disease, retinopathy, and neuropathy.

The long-term goals in living with diabetes are to reduce the symptoms and prevent diabetes-related complications, such as blindness, kidney failure, and foot or leg amputations. The ultimate goal, of course, is to prolong life and to make that life as comfortable and meaningful as possible.

11. Can you drown from drinking too much water?

On January 12, 2007, Jennifer Strange, a twenty-eight-year-old woman and mother of three children, was found dead by her mother in her Rancho Cordova home. She had taken part in radio station KDND's "Hold Your Wee for a Wii" contest to try to win a Nintendo Wii game console that went to the contestant who drank the most water without urinating. It is estimated that she drank two gallons of water in a very short period of time. Nearly all deaths related to water intoxication, excluding drowning, have occurred because of drinking contests, hazing incidents, marathon or Ironman races, and parents or babysitters punishing a kid.

Drinking too much water too quickly pushes the normal balance of electrolytes in the body outside safe limits. The level of sodium in the blood must stay within very narrow margins. Too low a concentration of sodium in the blood is called hyponatremia; other causes are not replacing electrolytes after exertion, large burns, severe diarrhea, heart failure, and certain medications. In addition to sodium, other electrolytes that must be kept with a reasonable balance are potassium, chloride, and bicarbonate.

Water enters the body orally and leaves the body in urine, sweat, and exhaled water vapor. If water enters the body more quickly than it can be removed, body fluids become diluted. As the body tries to balance the electrolyte concentration inside and outside its cells, osmosis makes the blood's concentration of electrolytes—mainly sodium and magnesium—drop relative to the concentration of electrolytes in the cells, causing the cells to swell. If this swelling occurs in the brain, pressure builds up because the bones that make up the skull don't budge. The brain squeezes against the inside of skull, which impairs brain function. The result may be headaches, nausea, impaired breathing, disorientation, muscle cramps, seizures, coma, and even death.

Drinking too much water also puts a heavy burden on the kidneys. They have to work overtime to filter out the excess water in the circulatory system. The kidneys don't work like plumbing pipes, which become cleaner the more water you flush through them; over time, unnecessary wear and tear can damage the specialized capillary bed in each kidney, called the glomerulus. The glomerulus is a network of capillaries that performs the first step of filtering blood, and serious deterioration can eventually lead to kidney failure.

The body is marvelous at letting us know how much water we need. Diet, exercise habits, and the environment all play a role. If you eat lots of foods naturally rich in water, such as vegetables, fruits, and whole grain cereals, you may not need to drink very much water. Avoiding salty foods will make you need less drinking water, too. Doctors say that the best guide to how much water we need is our sense of thirst.

12. Why does your heart stop when you sneeze?

This is just a myth; it only feels like your heart stops. A lot of pressure builds up in the chest, and that pressure spike may momentarily change the rhythm of your heartbeat, but it won't stop your heart.

A sneeze begins with a tickling in the nerve endings that tells the brain it has to get rid of something irritating the lining of the nose. A person takes a deep breath and holds it, the chest muscles tighten, the eyes close, the tongue presses up against the roof of the mouth, and "ah choo," the breath comes out the nose and mouth at speeds up to 100 mph. And it can spray up to five feet.

A heartbeat starts with a small bundle of tissue called the sinoatrial node, located in the upper part of a chamber of the heart called the right atrium. This natural pacemaker sends electrical impulses down to the right and left atria, then the ventricles to start that once-a-second heartbeat. Sneezing may affect the rhythm of the heart and may even cause the heart to "skip a beat," or throw the whole beat off, but in no way can it stop the heart. So sneeze away, but cover it up!

Why do people say "bless you" or "God bless you" ("*Gesundheit,*" meaning "health," in German, and "*salud,*" again meaning "health," or "*Jesús*" in Spanish) after someone sneezes? There are several origins of this custom.

People in some cultures once thought that when you sneeze, you open yourself, or your soul, to the outside world. And perhaps the soul might escape or demons or evil spirits might get into the body. The blessing is a way to ward off bad stuff.

Another theory of the origin of "God bless you" comes from the time the bubonic plague was sweeping through Europe. There was a lot of sneezing and coughing with the plague. Pope Gregory VII suggested that saying "God bless you" might protect a person from an otherwise certain death.

Whatever the origin, I think it is a very fine tradition and makes us a more civil people.

13. Why does a spinning motion cause nausea?

Every kid knows the joy of spinning around and around, getting dizzy, and trying to walk straight or simply stand up and not fall over, or perhaps of getting on one of those small merry-go-rounds at the park and rotating very fast and getting silly dizzy. What great fun!

This kind of dizziness goes to the heart of how we maintain our balance. There are three mechanisms at work to keep us from falling over. The primary tool for maintaining our balance is our vision. We can see if we are falling over and if things are level.

The second is our vestibular, or inner-ear, apparatus. The inner ear has three semicircular canals, all at right angles to one another and all filled with fluid. We're using the vestibular balance mechanism of our inner ear when we stand upright, close our eyes, and don't fall over. When a person spins around and around, they get that fluid moving. When the person suddenly stops spinning, the fluid keeps moving. There is conflicting information fed to the brain between what the eyes see and the messages that inner-ear fluid is sending. So a person feels dizzy for a bit. Astronaut and senator John Glenn had a fall in the bathroom while he was replacing a light fixture. He hit his head and had dizzy spells for some time due to damage to the inner ear.

The third component we use to maintain balance is the kinesthesia or proprioception mechanism. The two terms are often, and at times incorrectly, used interchangeably. Our body sends signals to the brain from our muscles and tendons, or proprioceptors, to give us an awareness of position and movement of our body. Aviators call it the "seat of the pants" sensation.

Dizziness or vertigo can also result from a medical problem; some drugs cause dizziness as a side effect, too. It's believed that 30 percent of Americans will visit a doctor sometime in their life complaining about vertigo, dizziness, and the accompanying nausea. The nausea is a result of conflicting messages sent to the brain.

In some people, especially older folks, dizziness causes falls and broken bones, including the rather serious hip break or fracture. There is the possibility that one's hip may break because of osteoporosis, and a fall may result. Many older people become dizzy when standing due to a sudden drop in blood pressure.

Astronaut Alan B. Shepard had severe vertigo after he contracted Ménière's disease, a disorder of the inner ear. It is caused by a build-up of fluid in the compartments of the inner ear, called the labyrinth. Ménière's disease can cause severe dizziness, ringing in the ear, and hearing loss. Shepard had such a severe case that he would lose his balance and fall. Shepard eventually got fixed up and flew on Apollo 14: He was the oldest astronaut to walk on the moon (1971), at age forty-seven, as well as the first American to fly in space (1961).

Dizziness, or vertigo, can be related to a neurological condition like multiple sclerosis or Parkinson's. It can really impact a person's independence, ability to work, and quality of life. It can lead to depression and disinterest in everyday life.

Doctors report that dizziness can sometimes be a difficult problem to analyze. It could require a battery of tests, including an MRI. The most common diagnosis is peripheral vestibular disease. This disease has a number of causes, from an otolith (stone) rolling around in one of those canals to "inner ear attack." An inner ear attack is an autoimmune condition that occurs when the body's immune system attacks the cells of the inner ear, mistaking them for viruses or bacteria.

14. What makes people's bones break, and how do they eventually heal?

Our bones are remarkable. Researchers at MIT unraveled the structure of bone with almost atom-by-atom precision. They found that bone is a complex mixture of collagen and mineral that forms into a composite to create a very strong, tough, and reliable material. Bones give our bodies structure, and they can bend, too. But if they bend too much, they break, or fracture. Compare bones to a wooden pencil. A wooden pencil will bend, but if you try to bend it too much, it will break. Bones behave the same way.

The average person will have one or two broken bones in their lifetime. The most commonly broken bone is the clavicle, or collarbone. The bones of the arm, wrist, hip, and ankle all are in the top five. And contrary to popular belief, there is no difference between a broken bone and a bone fracture.

A bone will break if pressure or an impact puts too much force on it. The most common causes are falls, car accidents, and sports injuries, all of which can cause a bone to snap like a wooden pencil or a tree branch.

Bones can also break if they are weakened from within, whether by natural aging or by osteoporosis, infections, or tumors. Bone is live tissue, a honeycomb of tough elastic fibers, minerals, marrow, and blood vessels, and it needs constant repair. Healthy bone is dense, with small honeycomb spaces. In older bones, the repair work slows, so the spaces get larger and the bone is less dense, weaker, and less elastic. In osteoporosis, a common condition in older people, the living bone cells are not able to break down old bone and replace it with denser and healthier new bone. Bone regeneration slows to the point where fewer new bone cells are being created than old ones are being lost. Over time, bones become thinner and are more likely to break.

The risk factors for broken bones include age, being female, low bone density, and a history of bone fractures.

Another cause of broken bones is repeated trauma to a bone over a period of time, to the point where the body can't keep up with repairing itself. Doctors use the term "stress fracture" to describe this condition; common examples include runners' ankle bones and hip bones.

But there is good news: Our bones are natural healers. At the site of the break, bones make a lot of new cells. New blood vessels develop that carry nutrients to help rebuild the bones. Those new cells cover both ends of the broken part of the bone and close up the break to make the bone as good as new, and possibly even stronger. Bones heal themselves by absorbing minerals and proteins from the bloodstream. The most important minerals are calcium and phosphorus, which are deposited in the bones to repair the break. Doctors put a broken bone in a cast to keep it immobile and aligned properly while all that depositing is taking place.

A new treatment for broken bones is a bone growth stimulator. The device, worn externally on the body, is used to treat nonunion breaks, where the broken bones have failed to heal. Electromagnetic or ultrasonic waves are delivered to the break site, which encourages bone growth.

There are different kinds of breaks, and X-rays are used to sort of map the break. A hairline fracture is a thin break in a bone. A complete fracture means the break goes all the way through the bone, leaving it in two pieces. A greenstick fracture is a crack on one side of a bone. A comminuted fracture is when the bone is broken into more than two pieces or is crushed. An open fracture is when the broken end of the bone sticks through the skin. Ouch!

How long does it take for bones to heal? Healing takes time, and the exact amount of time depends on the type of fracture and on the person's age, overall health, nutrition, and blood flow to the bone. A young child may have a bone heal in three weeks, but it may take six weeks for a teenager with the same kind of break. Bone material is created and deposited faster in young children than in teens and adults. What promotes faster healing for all age groups? Good nutrition, which means a balanced diet with all the food groups. Healing bones need more nutrients than intact bones that only need maintenance.

One of the worst things for healing bones is smoking. People who smoke have a much longer healing time and a higher risk of developing a nonunion, or nonhealing, of the bone. Smoking changes the blood flow to the bone, and it is that blood flow that delivers nutrients and cells to promote healing.

Doctors sometimes recommend calcium supplements, but there's a limit on how much calcium is helpful. The body will reject an oversupply of calcium. A more helpful recommendation than calcium overloading that your doctor can make is to follow a treatment plan, which may include rest, nutrition, and time, and, if need be, healing aids and procedures such as casts, crutches, surgery, or pins.

There is strong evidence that a lifetime daily intake of vitamin D is helpful for anyone who breaks a bone. Vitamin D, along with sunlight, promotes calcium absorption, which makes bones denser and stronger.

If you do not like the bones you have, just wait awhile. The average adult replaces their complement of skeletal bones about every ten years. It takes that amount of time to replace all the calcium in our bones. In fact, the skeleton is always regenerating and adjusting over the course of our lives. Bone remodeling alters our structure as our mechanical demands on our bodies change, and it helps make tiny patches of new bone to repair small damages.

15. Why do we get goose bumps when we are cold?

Goose bumps are small bumps on a person's skin at the base of body hairs. Tiny muscles at the base of each hair contract and pull the hair erect. The most common type of goose bumps is a response to cold. In addition to getting goose bumps, our bodies respond to cold in several other ways. Shivering increases the amount of heat produced by muscles by a factor of three or four times at peak shivering. As the body's core temperature drops, the central nervous system also restricts blood flow to the limbs and reroutes some of this blood to the internal organs of the body. The limbs get pale and then turn blue.

The animal kingdom also responds to cold. Birds preen and fluff up their feathers in anticipation of approaching cold weather. Fur coats thicken. Migrating animals go south. Some animals hibernate. Small animals eat almost constantly just to stay alive. The smaller the animal, the greater their surface-area-to-volume ratio. Small animals lose a lot more heat through their skin than big animals do. That is why there are very few small animals in arctic regions. Big animals, like polar bears, survive where squirrels would die. To me, the word "Arizona" sounds really good in the wintertime!

Goose bumps can also occur when a person is angry or afraid. Certain noises, such as nail scratches on a blackboard, can trigger goose bumps. Hearing awesome music, or seeing a famous person or someone we love, can produce goose bumps, too. I'm sure we've all heard people say something like, "That music was so beautiful it gave me goose bumps." Research published in the Canadian journal *Nature Neuroscience* indicates that dopamine is released when listeners are in a strong music-induced emotional state. The "reward" chemical dopamine produces a physical effect known as the "chills," which causes changes in heart rate, skin resistance, skin electrical conductance, and breathing rate.

16. Why do we have fingernails and toenails?

A fingernail is produced by living skin cells in the finger. The nail plate is the visible part of the nail. The nail bed is the skin beneath the nail plate. The cuticle is the tissue that overlaps the plate and rims the base of the nail. The nail fold is the skin that frames and supports the nail on three sides. The lunula is that whitish half-moon at the base of the nail, and the matrix is the hidden part of the nail under the cuticle.

Fingernails grow out of that matrix. Fingernails consist of keratin, the same hardened protein in hair and skin. New cells grow in the matrix and push out older cells, which become compacted and assume that familiar flattened, hardened form.

Fingernails grow all the time. They grow faster than toenails. Fingernails grow about three millimeters in a month. A millimeter is about the thickness of a dime. It takes about six months for the fingernail to grow its entire length, from the root to the free end. Growth slows down with age and with poor circulation. The growth rate also depends on the season and on a person's exercise level, diet, heredity, gender, and age. Fingernails grow faster in the summer, due to exposure to the Sun and an increase in vitamin D. But contrary to popular belief, fingernails do not continue to grow after we die. The skin around the fingernails dehydrates and tightens, making the nails appear to grow.

There are a number of theories about why we have fingernails and toenails. The most common relates to how humans altered or adjusted to a changing environment over many eons. Fingernails and toenails are almost exclusively a feature of humans and primates. Whatever the reason for their existence, fingernails are certainly useful. Fingers are very sensitive; each contains a mass of nerves. Fingernails protect our fingertips. They also make us more dexterous and help our hands do amazing things. Fingernails help us scratch when we itch. They help us peel things like oranges. Fingernails help us undo knots in string and ropes. They help us grasp or pick up things that are very tiny, such as a nut, screw, needle, peanut, or pencil. We use fingernails to grip, rip, and tear. People who have lost fingernails, or toenails, for that matter, can attest to how valuable they are.

Healers have used the general appearance of fingernails as a diagnostic tool since olden times. Some major illnesses can cause a deep groove to form across the nail. Brittleness, splitting, discoloration, thinning, white spots, and Mees' lines can indicate problems in other parts of the body. Mees' lines, which are white bands across the width of the nail, appear when a person is poisoned with arsenic, thallium, or another heavy metal. Mees' lines show up in some patients undergoing chemotherapy, and kidney failure can also cause their appearance.

17. What are people made of?

There are several ways to measure what we are made of. We can classify ourselves as amounts of fat, bones, and muscle. Another way we can appraise ourselves is by percentage of water. About 50 to 60 percent of our bodies is water. It's a good thing we are warm-blooded; otherwise, we might freeze to death!

At the most basic level, people are made of the same stuff that everything else is made of—namely, atoms. The atoms make up molecules, and the molecules make up chemicals. We can also classify ourselves by the chemicals in us. If we go by chemical composition, we are mostly carbon. We are about 62 percent carbon, 11 percent nitrogen, 10 percent oxygen, 6 percent hydrogen, 5 percent calcium, 3 percent phosphorus, and 1 percent potassium. That gets us up to about 98 percent. The other 2 percent is trace amounts of about twenty-eight other elements.

Yet another way we measure ourselves is by fat content. According to the National Institutes of Health, a healthy adult man should have a body fat of between 13 and 17 percent. A healthy adult woman should have a body fat of between 20 and 25 percent. The ideal percentage of muscle is about 43 percent.

We can compare the efficiency— defined as work done divided by energy consumed—of the human body with the efficiency of other machines. A human can get up to about 20 percent efficiency by cycling on a stationary bike. The top efficiency of a gasoline engine is about 38 percent, but most are around 20 percent. So we are as good as most engines.

18. How does your heart pump?

The heart is a fist-size, muscular organ that pumps blood. And it is very busy, pumping five quarts of blood every minute, or almost eighteen hundred gallons a day.

The heart has the job of pushing blood through our arteries, which carry oxygen and nutrients to all parts of our body. Blood returns to the heart via veins called the superior and inferior venae cavae to pick up a fresh supply of oxygen and nutrients.

The heart has four chambers. The top two chambers are atria, and the bottom two are ventricles. The right side of the heart pumps blood to the lungs, and the left side pumps blood to the rest of the body. Each chamber has a one-way valve to keep any blood from flowing backward. The top two atria are receiving chambers. The right atrium receives blood from the upper part of the body via the superior vena cava, and blood from the lower extremities by way of the inferior vena cava.

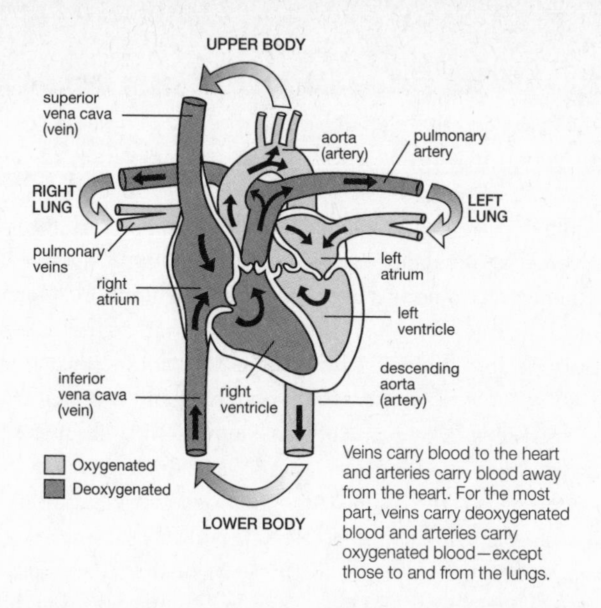

Veins carry blood to the heart and arteries carry blood away from the heart. For the most part, veins carry deoxygenated blood and arteries carry oxygenated blood—except those to and from the lungs.

The left atrium receives blood from the lungs. The bottom two ventricles are pumpers, the right pumping blood to the lungs and the left pushing blood to the other organs and tissues. That left ventricle does about 80 percent of the work of the heart. The atria and ventricles work together, alternately squeezing (contracting) and then relaxing. The heartbeat is triggered by electrical impulses that originate in the sinoatrial node, located in the right atrium. Those electrical signals are conducted across both atria by nerve fibers. This makes the atria contract and squeeze blood into the ventricles. The signals from the atrium on each side cross a junction to the ventricle on that side, telling the ventricles to contract and pump blood to all parts of the body.

We are all one heartbeat away from eternity, so it pays to take care of such a vital organ. That means, of course, having a heart-friendly diet that is low in fat, low in salt, and low in sugar (see page 8). And keeping our body weight in check is not always easy, but it is important: Being overweight or obese puts people at high risk of heart disease, which is the leading cause of death in the United States. The toll is over six hundred thousand per year.

We should all get regular aerobic exercise designed to improve our oxygen consumption and raise our heart rate. That could be such activities as walking, jogging, biking, and swimming. There also seems to be a genetic component to having a healthy heart. So be thankful if you have "good heart genes."

19. Why does blood in our veins look blue?

When I was a kid growing up on a farm, we kids would argue about this. The general consensus was that blood was blue when it was in our body but would turn red when it hit the air. Well, we were mostly wrong.

It turns out that blood is never blue. Its color—either red or darker red—is determined by the amount of oxygen and carbon dioxide in the blood. We know that blood contains hemoglobin, which contains iron atoms (see page 7). (Rock formations with a high iron content are reddish in color.) In the lungs, there is a lot of oxygen available, which bonds with the hemoglobin. So blood leaving the lungs through the pulmonary veins—the only veins to carry oxygenated rather than deoxygenated blood—returns to the heart high in oxygen content and bright red in color. The heart pumps this red blood to the rest of the body through arteries to deliver oxygen to tissues, organs, and muscles, where it is used up (see page 27).

The blood returning in a separate set of veins is depleted of oxygen, which has been replaced with plenty of carbon dioxide, giving the blood a much darker red color. And when we look at our veins, the color of the blood appears bluish because some of the dark red color is absorbed by the veins and skin. (This only works for veins because we can't actually see our arteries through our skin; they have muscular walls that are much thicker than our more thin-walled veins.) Mostly blue color is transmitted.

I used to show the skin-absorption phenomenon to students in science class by taking a small glass tube and filling it with water made bright red using food dye. Both ends were capped with clay. I would place the "blood-filled" glass tube in a glass tray and slowly pour skim milk into the tray. The color of the "blood" seemed to slowly change from a reddish to a bluish tint as the milk covered the glass tube. The milk absorbed some of the longer red wavelengths of light coming from the tube of "blood." The shorter blue wavelengths of light were transmitted through the milk, yielding a bluish tint to the "blood" tube.

20. Why is chicken pox so much worse for adults than it is for kids?

The cause of chicken pox is a virus called varicella-zoster. This highly contagious disease spreads from person to person by direct contact, coughing, or sneezing.

The incubation period is between ten days and three weeks. Most of us who were children before 1995, when the vaccine became widely used, remember the red, itchy rashes and blisters on our skin. And, of course, we remember staying home from school for a few days. Once a child gets over chicken pox, they have lifelong immunity. They never get it again.

Treatments include cool, wet compresses or baths, calamine lotion, and acetaminophen, such as Tylenol. Aspirin should be avoided. It can cause Reye's syndrome, a potentially fatal disease that has a detrimental effect on many organs, including the brain and liver. Ibuprofen, the medicine in Advil, may cause more severe secondary infections and should also be avoided.

Getting the chicken pox (varicella) vaccine can protect children from the chicken pox virus. The vaccine reduces the chances of getting chicken pox. Vaccinated kids who do get it will have a much milder case.

Kids, in most but not all states, must be vaccinated for chicken pox before they enter preschool or kindergarten. Typically, kids receive the vaccine between the ages of twelve and eighteen months, plus a booster before starting kindergarten. A teenager who has not received an inoculation for chicken pox has an increased risk of contracting it in adulthood, and chicken pox is harder on adults than it is on children. Adults tend to get sicker. Stacey Rizza, chair of the HIV clinic at Mayo Clinic, states that "getting chicken pox as an adult can lead to more severe symptoms because your immune system is not as young and ready to attack, so you get a little sicker." There is also the added danger of pneumonia. Adult inoculation against chicken pox may require two shots.

Shingles is a second eruption of the same varicella-zoster virus that causes chicken pox. And shingles can be bad news. After an attack of chicken pox, the nasty virus can hide among nerve cells of the ganglia and spinal cord. A ganglion is a biological tissue mass, most commonly a mass of nerve cell bodies.

Because the virus may never have fully disappeared from the body, it may lie dormant for years. Doctors do not know why the virus flares up suddenly after years of inactivity. Disease, stress, age, or a weakened immune system can cause the virus to reactivate and reproduce. It travels along the path of a nerve to the skin's surface, where it causes shingles. Shingles is characterized by burning pain, sensitive skin, rash, and blisters. The blisters pop and ooze. The whole process unfolds over three to four weeks. Antiviral drugs, such as Zovirax, Famvir, or Valtrex, are prescribed for treatment.

21. Why do some people find it hard to remember things after they have had a head injury?

There are different types of memory, short-term and long-term, and the brain stores them in different places. Short-term memory is good for just a few minutes. If someone gives you a seven-digit telephone number to call, you store that information in short-term memory. Most of us will have forgotten it an hour later. The storage of information in short-term memory is limited in capacity and time. Long-term memory stores things you remember for a year, five years, ten years, or a lifetime. If short-term memories are practiced often enough, they turn into long-term memories. That's the reason most people know their Social Security number "by heart"; they've written it down or "practiced" it often enough.

The brain's hippocampus is essential to taking information from short-term memory and putting it into long-term memory. Some scientists believe that long-term memories are maintained by stable and permanent neural connections widely spread throughout the whole brain. One of the main functions of sleep is the consolidation of information. The hippocampus replays information from that day and speeds it along to long-term memory.

Memory loss is the most common cognitive impairment in people with severe head injuries, including some loss of specific memories and partial inability to form or store new ones. Their long-term memory tends to be all right. They will say to their doctor, "I can remember something that happened ten years ago in great detail, but I can't recall something from ten minutes ago."

It is common for brain-injured people to forget events that occur right before, during, and right after the injury. This temporary loss is caused by the brain's swelling, which presses the brain against the inside of the skull. Memory usually returns when the swelling goes down.

So why do these injuries affect short-term memory? It has to do with how the brain processes information. Info from our senses goes through a filtering process, sort of like how mail is sorted at the post office. When the brain is injured, the areas that do the processing can get pressed or squeezed due to swelling. When this happens, a large amount of information and sensory data coming into the brain does not get processed. The information is not sent to the right places. The mail room of the brain can't do its job.

The brain can also have another type of memory problem. Once data is stored in the brain, it has to be able to retrieve it. But the brain can have a problem retrieving stored information. We've all had this type of problem, but in the brain-injured person it is much worse. You meet someone on the street. You know who they are but can't quite come up with their name. A few minutes or hours later, you recall the name. The brain has been searching and trying to retrieve that bit of information. For people with brain damage, retrieval of information may be difficult or impossible because the connections have been permanently lost.

Most brain injuries are caused by car accidents, motorcycle and bicycle accidents, and falls around the house. But recently we are hearing about head injuries to our soldiers who have served in Iraq and Afghanistan from those improvised explosive devices (IEDs). Also, there is now a lot of concern about brain damage in football, where head-to-head collisions can cause concussions, confusion, drowsiness, nausea, disorientation, lack of coordination, and slurred speech, in addition to long-term effects. The NFL is looking into this matter. There is also much concern about possible brain damage to high school and college football players.

Recent research indicates that a third type of memory, called immediate memory, may be at play in the brain. Immediate memory is a blanket term that may include short-term memory and there is overlap between the two.

Most researchers agree that the study of the human mind is in its infancy. It is an exciting area of study, and there is much to learn.

22. How many organs are there in a person's body?

There are ten or eleven main organs and a whole bunch of little ones. The exact number is hard to come up with, because there are different views on what qualifies as an organ. Do ligaments and tendons count as organs? It's difficult to decide which are organs and which are just parts of a larger system.

An organ is defined as a structure that contains a collection of tissues working together for a common purpose. Medical dictionaries also define an organ as a relatively independent part of the body that carries out one or more special functions.

The main organs are the brain, heart, lungs, liver, kidneys, stomach, bowel, intestines, and skin. The skin is the largest organ. It accounts for about 15 percent of our body weight. The skin protects everything that lies beneath it. It alerts our body to dangers and acts as a cushion to blows. Our skin is a barrier and the first line of defense against invading parasites and diseases.

The largest internal organ is the liver. It is also the heaviest, weighing in at about 3.5 pounds. The liver regulates most chemical levels in the blood and excretes bile, which helps break down fats. The longest and strongest bone is the femur, the long upper leg bone in the thigh. The largest artery is the aorta, the big one that sits atop the heart and carries oxygenated blood to all parts of the body, while the largest vein is the inferior vena cava, which returns blood from the bottom half of the body back to the heart.

Biology and anatomy students have used mnemonic, or memory, devices for years to remember material that just might show up on tests. A few examples: GRIM END is an explanation of the seven aspects of life: Growth, Reproduction, Irritability, Movement, Excretion, Nutrition, and Death. MRS GREN for the seven things all living animals do or have. Movement, Reproduction, Sensitivity, Growth, Respiration, Excretion, Nutrition. I Picked My Apples Today is a mnemonic for remembering the phases of mitosis, the process of cell division, which allows our cells to multiply into all the organ systems of our bodies: Interphase, Prophase, Metaphase, Anaphase, Telophase.

23. How do you get hiccups?

Most people are more interested in how to stop hiccups! Hiccups are unintentional movements, or spasms, of the diaphragm followed by rapid closing of the vocal cords. A spasm is a sudden muscle jerk. The diaphragm is a thin dome-shaped muscle at the bottom of the chest. It separates the organs in the chest (heart and lungs) from the organs in the abdomen (stomach, liver, spleen, pancreas, gallbladder, and intestines). When a person inhales, it is because the diaphragm is pulling down to help pull air into the lungs. When a person exhales, it is because the diaphragm is pushing up to help push air out of the lungs. Intercostal muscles, which lie between the ribs and have a lesser role in breathing, are also affected by hiccups. These muscles help us breathe, speak, sing, and cough.

When the diaphragm becomes irritated, it misbehaves. It pulls down in a jerky way, sucking air into the throat suddenly. When that rushing air hits the voice box, it makes the sound we know as a hiccup. The causes are different for each person. But one thing is common to all hiccup cases: irritation of the diaphragm. Hiccups can be caused by eating too much or too fast. Drinking very hot or very cold beverages can also bring on hiccups. Other causes include cold showers, entering or leaving a hot or cold room, or any sudden excitement or stress. Eating spicy foods, drinking alcohol, breathing in suddenly, sneezing, laughing, and coughing can all lead to hiccups. Sometimes the cause is unclear; very often, hiccups start for no apparent reason. Sometimes when a woman is pregnant, the fetus even gets hiccups inside her!

Hiccups are not a serious problem; they're just an annoyance. They usually go away by themselves in a few minutes. Just as there are many causes, there are numerous methods people use to stop hiccups. A favorite cure is to drink a glass of water. Another popular method is to stretch the diaphragm by holding one's breath while raising one's arms. Anything that promotes build-up of carbon dioxide in the blood will stop hiccups.

Distractions, such as saying the alphabet in reverse or concentrating on a painting, seem to work for some people. A half-teaspoon of dry sugar or honey might work, too. If the vagus nerve that runs from the brain to the stomach is stimulated, hiccups can be relieved. Having someone frighten you or startle you may work. A tug on the tongue works for some people.

Children tend to "grow out" of having hiccups by the late teen years. While most people get hiccups from time to time, the frequency of hiccup episodes decreases with age and maturity.

In extreme cases, people might have to see a doctor. A doctor can interrupt the hiccups by giving muscle relaxants, sedatives, or anticonvulsive drugs.

24. How did different skin colors come to be?

Skin color result from the presence of a pigment called melanin, located in the epidermis, or outer skin layer, and produced by cells called melanocytes (see page 6). Melanin acts as a protective shield against ultraviolet radiation and helps prevent sunburn damage that can lead to melanoma (skin cancer).

There are several theories concerning the evolution of skin color. A long-held theory suggested that people in the tropic regions near the equator developed dark skin to block out the stronger Sun. The weaker the ultraviolet light, the lighter the skin. The greater levels of melanin in dark-skinned people would prevent them from overdosing on vitamin D, which can be toxic in high concentrations. However, critics of this theory say that it is impossible to overdose on natural levels of vitamin D.

Humans need some ultraviolet rays to penetrate the skin in order to produce vitamin D, which is needed for the intestines to absorb calcium and phosphorus from food and deposit it in our bones. Lack of vitamin D results in rickets in children and osteoporosis in adults (see page 57). People living above fifty degrees north latitude or below fifty degrees south latitude have a much higher risk of vitamin D deficiency. Only the ability to fish, which provides access to food rich in vitamin D, allowed early humans to live in temperate and polar climates.

A newer theory proposes that differences in skin color are related to the fact that ultraviolet light from the Sun affects the level of folic acid (a type of vitamin B) in the body. An hour of intense sunlight can cut a light-skinned person's folic acid level in half. People who live in the tropics developed dark skin to block out the Sun and protect their folate reserves. People who live farther north or south evolved to have fairer skin so they could absorb more ultraviolet light from the Sun and produce needed vitamin D, but they also need to get more of their folic acid from their diet. Fair-skinned people also have a higher risk of skin cancer. Low levels of folic acid are associated with neural tube defects, such as spina bifida. For this reason, women in the early stages of pregnancy should avoid tanning booths. Women of child-bearing age should make sure to get enough folate in their diet even before they become pregnant.

Some anthropologists believe that both theories may be correct. They believe people close to the equator developed darker skin to block out the Sun and protect their body's folic acid reserves. People closer to the polar reaches developed lighter skin to produce adequate amounts of vitamin D during the longer winter months.

25. What makes someone tone-deaf?

The technical term for tone-deafness is amusia, and one in twenty people have it. Tone-deafness is the inability to distinguish between musical notes. To clarify definitions, "tone," "pitch," and "frequency" all mean the same thing, namely the number of vibrations per second. Different frequencies of vibrations create different sound waves, which we perceive as different notes. There is not much correlation between a good singing voice and the ability to hear tones accurately. Some people who have bad singing voices hear music just fine. However, the inability to keep time with music (or lack of rhythm) and the inability to recognize common songs are indications of tone-deafness.

There has been some recent and intriguing research on this subject. New brain-imaging techniques can measure the density of the white matter, consisting of nerve fibers, that provides paths between the right frontal lobe and the right temporal lobe. The right frontal lobe is where higher thinking occurs, and the right temporal lobe is where sound processing takes place. In tone-deaf people (amusics), the white matter connecting them is thinner, which weakens the connection between these two parts of the brain. Studies show a direct correlation: the thinner the white matter, the worse the tone-deafness. For a tone-deaf person, the neural highway is a dirt road and not a four-lane interstate!

Some believe there is an overlap between how the brain handles music and how it handles speech. Music involves the whole brain. Other researchers believe that musical perception and thinking occur separately from other functions. Many scientists say that there is a strong genetic component to tone-deafness.

You can go online and check your ability to perceive tones at delosis.com/listening/home.html. The site is from the Harvard Medical School.

Charles Darwin, General Ulysses S. Grant, President Theodore Roosevelt, and William Butler Yeats were all tone-deaf. These men achieved greatness despite their tone-deafness.

26. What is human hair made of?

Human hair is made of the same stuff that horns, fingernails, cat and dog claws, cow and horse hooves, and bird feathers are made of: a protein called keratin. And it grows out of a tiny opening in the skin called a follicle. One kind of keratin is made of long protein chains, and the coiled chains wind around each other to resemble a shape much like a twisted telephone cord.

Keratin was first described by Linus Pauling, who won two Nobel Prizes; one was in chemistry, for applying quantum physics to chemical bonding in chemistry, and the other was the Nobel Peace Prize, for his opposition to the spread of nuclear weapons.

All parts of the body have hair except the lips, palms, and soles of the feet. Adults have about five million hairs growing out of the body. That's about the same number as a gorilla. Thank goodness, human body hair is thin and short and hard to see. Gorilla hair is thick and long and, well, hairy.

It's widely held, though not entirely proven, that about eighty hairs fall out of our head each day. Not to worry. Our head has over one hundred thousand hair follicles. Plenty to spare.

Some people have curly hair and some have straight hair. It depends on the shape of the follicle out of which the hair is growing. Straight hair grows out of a round follicle. Curly hair comes out of an oval follicle; it bends because its cross section is oval, not round.

Hair color is caused by the same chemical pigment that determines skin color—namely, melanin (see page 34). White hair means no melanin. Lots of melanin means black hair. Smaller amounts of melanin mean blond, red, or brown hair.

Hair's function is to keep us warm. When we are chilled, we get goose bumps (see page 23). Those temporary bumps are caused by muscles attached to hair follicles pulling the hairs upright.

I've been watching the hair fall on the cloth when I go to the barber. Still a lot of brown and dark-brown color there, but I do believe I notice some hairs with a grayish tint to them!

27. Why do we age?

Things wear out, and our bodies are no exception. Aging is a natural occurrence and not a disease. There seem to be two aspects to aging: nature (genetic influences) and nurture (environmental influences). Aging, to a large extent, is under genetic control. Many scientists say that we are genetically programmed to die in less than one hundred years.

Individual cells of our body have different life spans. Stomach cells last about two days, red blood cells 120 days, bone cells about thirty years, and some brain cells a lifetime. But while these cells reproduce before they die, the organism as a whole has a limited life span. No matter how well we take care of ourselves, our bodies wear out, shut down, and die.

The other aspect to aging, the nurture, or environmental, part, is largely the accumulation of changes, or mutations, in our DNA. These lead to loss of metabolic capacity; muscle and skin cells slowly lose their ability to regenerate with time.

The good news is that we do have some control of the aging process. Good diet, moderate use of alcohol, avoidance of tobacco, avoidance of overexposure to the Sun's ultraviolet rays, and an exercise regimen all contribute to slowing the aging process.

How long do things last? A sequoia tree will live for 2500 years. A mouse typically lives fewer than four years. A housefly has a life span of twenty-five days. A facelift is good for ten years. A pencil will write 45,000–50,000 words. A dollar bill is in circulation for eighteen months.

As for us, American men live about seventy-four years and women around seventy-nine years on average. The leading causes of death in the United States are heart disease (1 out of 3) and cancer (1 out of 4). Many dread diseases, such as polio, typhoid, smallpox, and diphtheria, have been conquered, so accidents are now the leading cause of death for people fifteen to twenty-four years old.

Some scientists believe we may one day unlock the secret of immortality, but do we really want people to live forever? Think of the overpopulation. Think of the lines at McDonald's!

28. What makes our eyes twitch?

Eye twitches are harmless, involuntary spasms of the tiny muscles surrounding the eye. Eye twitching comes from squinting too much, or simple fatigue, or drinking too much coffee, or dry eyes, or by staring at a computer monitor or television set for too long a period of time.

People who spend a good part of the day at a computer screen are prone to eye twitches. Spasms occur with overuse of any muscle in the body. So simply resting the eyes will take care of most eye twitching. Blinking now and then helps.

Most eyelid twitching disappears without any treatment. But, if you want to stop the twitching dead in its tracks, try rinsing the eyes with warm water, administering antihistamine eyedrops, applying warm compresses, or taking a warm bath. Minor eye twitching, the kind that most all of us experience, usually does not worsen. Some people have eye twitches on one side of the face and not the other.

Blepharospasm is a medical term for the twitching of one or both eyelids, commonly associated with stress. These spasms can last a day or so, and then they usually disappear. When they don't, the medical profession has used Botox injections to relax the surrounding muscles. Botox is the preferred treatment for most sufferers.

Blepharospasm can be one of the side effects of medications used to treat epilepsy and psychosis. Severe types of blepharospasm can be attributed to dry eyes, Tourette's syndrome, or neurological problems, such as epilepsy. These severe types require a doctor's care.

29. How does our body prepare for cold weather?

Our bodies have wonderful defense mechanisms against threatening, low temperatures. When we are exposed to cold temperatures, the pilomotor reflex triggers an involuntary muscle contraction that raises the hairs on our skin, producing goose bumps. This traps air near the skin, thereby retaining heat for the body. When we shiver, muscles contract and relax quickly, producing heat. Shivering continues until enough heat is produced. Shivering increases body heat being generated by our muscles many times over. Finally, the brain goes into a survival mode by sending messages that restrict blood flow to the limbs. Blood is shunted to the internal organs. Arms and legs get pale and turn blue.

How cold can we get and still live? Hypothermia sets in below a body temperature of ninety degrees; we lose the ability to shiver and speech becomes slurred. At around eighty-six degrees, we become unconscious, sink into a coma, and become rigid.

Staying dry is important. We lose twenty-five times more body heat if we become wet, such as when we break through ice and fall into the water. Warren Churchill, fishery biologist with the Wisconsin Department of Natural Resources, is said to be the "coldest man alive." In April 1973, Churchill fell into chilly Lake Wingra near Madison. His core temperature went down to sixty-one degrees. Doctors put him in a special blanket that had warm water circulating in tiny tubes. Churchill shivered so badly doctors had to inject a drug that paralyzed his body. At such extreme temperatures, shivering is so hard on the body that muscles can be strained and torn.

30. Why does an extra chromosome cause Down syndrome?

There are several types of mental defects that involve an extra chromosome. The most common is Down syndrome. In 1866, an English doctor by the name of John Langdon Down published an essay in which he described a set of children with distinct facial features.

There are different types of Down syndrome. Down syndrome is caused by abnormal cell division, most often in the woman's egg before or at the time of conception. Children inherit forty-six chromosomes, twenty-three from each parent. Genes grouped together make up chromosomes. Chromosomes carry the genetic material DNA, or genes. One of the twenty-three pairs of chromosomes, called X and Y, determines the sex of the child. DNA in other chromosomes determines such things as blood type, hair and eye color, and risks for certain diseases.

Defects in chromosomes can cause changes in body processes and functions. Sometimes these changes are undetectable. Genetic defects can pass from parent to child or can occur because of a new mutation. The genetic cause of Down syndrome is full trisomy 21, with a whole extra twenty-first chromosome; in rare cases, Down syndrome can also be caused by inheriting merely some of this chromosome's genes. Down syndrome occurs when there is an error in cell division. In 95 percent of the cases, one cell has two copies of chromosome 21 instead of one, so the resulting fertilized egg has three copies of chromosome 21 instead of two, hence the name trisomy 21.

Some general characteristics of Down syndrome include reduced height, poor muscle tone, broad neck, shorter-than-normal arms and legs, protruding abdomen, flat nasal bridge, and abnormal shape to the ears.

Severe mental retardation is rare, but most Down syndrome children have reduced intelligence ranging from mild to moderate. Some Down syndrome children have heart defects or vision problems, and they may suffer from hypothyroidism and respiratory problems. Women over thirty-five have an increased risk of having a child with Down syndrome. The risk increases with advancing age.

Life is precious, and with love, care, and education, most children with Down syndrome can lead productive lives.

31. How do we grow?

Many living things, including us humans, grow by cell division. Each of our cells has a nucleus, and during a process we call mitosis, the nucleus divides, creating two cells from one. Each new cell receives a copy of the parent cell's genetic material.

Cells are forever dividing, creating more cells and leading to tissue and bone growth. Organs grow, skin grows, and everything in the body just keeps getting bigger through childhood. The human body grows constantly and steadily from birth to about age eighteen or twenty. You are not likely to get any taller after age twenty. Of course, there is the possibility of getting wider. Bone growth—in leg bones in particular, because they grow on both ends—is what determines how tall we are (see page 51). Bones get longer, but not much wider, until we get to our late teens, when bones generally stop growing altogether.

Bone growth and height depend on your genes to a large extent, and somewhat on your diet. Bones need calcium and vitamins. So drink milk and skip the sugary drinks, or at minimum, limit your intake. Muscles grow along with the bones, and most of their growth is automatic, although some muscle growth is only triggered by exercise. Use it or lose it!

The pelvis widens one inch between the ages of twenty and eighty, even if a person watches their weight and keeps the same level of body fat. That means about a three-inch increase in waist size. Sometimes life is not fair!

The skull also continues to grow larger as we age, and the forehead shifts forward a bit, making the cheekbones recede slightly. Growth rate is not uniform for all body parts. When a baby is born, its head size is nearly that of an adult's, but the lower parts of the body are much smaller. Then, as a child develops, the head grows very little but the legs, arms, and torso increase a great deal in size. Heredity determines growth rate to a great extent, but nutrition, exercise, injury, and disease are all factors.

While body growth is so complicated that some of its workings elude us to this day, again, everything comes back to one thing: cell division.

32. How do our brains work?

The brain has been referred to as gray matter since at least 1840. Truth be told, the brain is a pinkish, fleshy color. The very center of the brain is an off-white shade. The brain is very soft tissue, having the consistency of pudding, and it weighs about three pounds and has a volume of 1300 cubic centimeters (about 18.3 cubic inches). Brain weight and size vary with the size of the individual.

Each of the 100 billion neurons in the brain has about seven thousand connections to other neurons, creating a huge network of more than 100 trillion synapses. Each connection is "on" or "off," like transistors in a mega-computer.

The brain uses about 20 percent of our total oxygen intake and about 25 percent of the glucose we consume. The oxygen is used to get energy from the glucose, which is the brain's source of energy. If the brain's oxygen is cut off, permanent brain damage occurs after four minutes. Hypoxia means low on oxygen and anoxia means total lack of oxygen.

There are three main parts to the brain: the cortex, the limbic system, and the brain stem. The cortex handles the most complicated things, such as thinking, making decisions, and recognizing sights, words, sounds, and sensations. We depend on the cortex for playing sports and music and for writing. The limbic system is involved with survival. It lets us know when we need to eat, drink water, and put on a coat. The limbic system warns us of dangers and makes us aware of threats, and it's where we experience pleasure and happiness. The brain stem connects the brain to the spinal cord, which runs down through the backbone. The brain stem controls heart rate, breathing, and other vital functions. If it is badly damaged, a person can lose consciousness and lapse into a coma. The cortex needs the brain stem to keep it alive.

A whole slew of things can go wrong with the brain. Heart attack, suffocation, drowning, high altitude, and head injury can all put a damper on a healthy brain.

There are two kinds of strokes both caused by interruption of blood flow to part of the brain. A blood clot formed in a blood vessel can break off and block an artery in the brain, causing a thrombotic stroke. An aneurysm happens when an artery wall is weakened. The damaged area can swell and apply undue pressure to the surrounding tissue, or burst, causing uncontrolled bleeding and disrupting the brain's blood supply; this is called a hemorrhagic stroke. Like most of the body's tissues, the brain can develop tumors, growths caused by runaway cell division. Malignant, or cancerous, tumors invade surrounding tissue and can caused massive damage or spread to other parts of the body. Benign, or noncancerous, tumors do not spread or attack other tissue, but they can grow quite large, putting pressure on adjacent brain tissue.

The abuse or misuse of legal and illegal drugs can damage nerve cells in the brain and lead to permanent brain damage.

Dementia is a general term that describes a wide range of brain declines, such as memory loss, deterioration of thinking skills, and the inability to perform everyday activities. Alzheimer's accounts for about 50 to 70 percent of cases.

Sometimes a medical examiner or coroner will order an autopsy of a body (see page 72). The reason, of course, is to establish cause of death. As part of most autopsies, the brain is removed. The medical examiner uses an electric saw, called a Stryker saw, to make a round cut through the top of the skull. The cap of skull bone is removed. The medical examiner employs a scalpel to cut the tissue that connects the brain stem to the spinal cord. The brain can be pulled out and stored in a solution to keep it available for further examination.

Albert Einstein's brain was removed within a few hours of his death in April 1955. It is well worth reading about the journey his brain took the last nearly 50 years—try Carolyn Abraham's *Possessing Genius*.

The brain is a wonderful instrument, and it makes us who we are; our bodies are just along for the ride, so to speak, and, when separated from the ingenuity of our brains, are quite utilitarian. The brain is so complex, it has been referred to as one of the last frontiers of the unknown, alongside outer space and the deep ocean.

Our brain is so magnificent and exquisite that it behooves us to take good care of it, for no other reason than it is the only one we will ever have. Realize that it is not wise to endanger our brain by drug or alcohol misuse or by failure to wear bike helmets or seat belts. And we know that, like our muscles, our brain needs exercise—in this case, by lifelong learning, although physical exercise benefits our brain, too.

33. Is there anything we can take to stop a heart attack?

A person experiencing a heart attack or suffering from angina can take nitro pills, short for nitroglycerin pills, which rapidly open blood vessels. Angina pectoris is chest pain that occurs when there is not enough blood flow to the heart. The medication dilates, or opens, the coronary arteries to the heart, which reduces the burden on the heart, and the pain.

Nitroglycerin comes in a fast-acting tablet or in spray form. Both are applied under the tongue. A sufferer can use the tablets for fast relief in addition to wearing a longer-acting nitro patch. Nitroglycerin tablets are usually taken every five minutes, up to a total of three tablets. If the person is still having heart pain, they had better go to the emergency room and see the doctor immediately.

The nitro pills do come with some warnings. Headache, dizziness, and light-headedness are common side effects. We see all those erectile dysfunction drugs (Viagra, Levitra, Cialis) advertised on TV. The announcer always reminds the viewer (in a low and fast-talking voice) not to combine these drugs with nitroglycerin medications. That might lower blood pressure to a dangerous, life-threatening level.

Mucosa, or mucous membranes, is the moist tissue that lines particular organs and body cavities throughout your body, including your nose, mouth, lungs, and gastrointestinal tract. Glands along the mucosa secrete mucus (a thick fluid). Placed beneath the tongue, nitro is absorbed directly through the mucosa into the bloodstream. With this method, the nitroglycerin is absorbed within a few seconds, faster than when the nitro pill is swallowed, digested, and absorbed.

Aspirin is a well-known treatment for preventing heart attacks, too. Aspirin thins the blood by reducing the clumping action of platelets.

Many doctors will recommend a daily regimen of low dosage aspirin, usually 81 mg, for their heart patients.

34. How does anesthesia work?

One of the benefits of writing a science column is that I have an excuse to look up a lot of fascinating stuff. It makes me realize how much I have to learn about so many different topics. It also gives me an opportunity talk to people who are experts in their chosen profession. To make sure I answered this question exactly right, I did some of my own research and talked to a real, live anesthesiologist.

On October 16, 1846, William Morton, a Boston dentist, used an ether-soaked sponge to sedate a printer named Gilbert Abbott. Then Dr. John Collins Warren removed a vascular tumor from the patient's neck. The patient later informed the onlookers that he had experienced no pain at all. Several men claimed to be the first to make use of anesthesia, including Crawford Long, Horace Wells, and Charles Jackson. Jackson also takes credit for the telegraph, an invention we attribute to Samuel Morse. A new era of medicine had begun.

Today, there are four basic types of anesthesia: general, regional, and local anesthesia and sedation. Each type has an effect on a different part of the nervous system. General anesthesia affects brain cells and causes a person to lose consciousness. Regional anesthesia is administered at different levels of the nervous system to block nerves, while local anesthesia is for a small, specific area. Regional anesthesia may include spinal blocks, epidural blocks, or nerve blocks. Nerve blocks may affect an arm or leg that is being operated on by the surgeon.

Sedation is like twilight sleep. Some of the drugs used for general anesthesia can be used for sedation, but at a lower dose. Examples would include novocaine employed by the dentist or a topical anesthesia used to numb the skin surface. (Sedated is kind of like the state of those soon-to-graduate seniors at the high school. Ha, ha . . . a little joke.)

General anesthesia works on the cerebral cortex of the brain, the thalamus, and the brain stem, resulting in immobility. It also affects the brain stem's reticular activating system (RAS), which leads to unconsciousness. The RAS regulates our sleep-wake transitions. So a person under proper general anesthesia is pain-free, immobile, and unaware of what is happening to them. When the patient wakes up, they have no memory of the period of time under anesthesia.

I recently had eye surgery using general anesthesia. I always wondered how the doctor knew how much anesthesia to give a person: too much and I go to that "great classroom in the sky," not enough and I wake up during the operation. And I wasn't keen on either condition. So, the anesthesiologist told me, he monitors heart rate, heart rhythm, blood pressure, breathing rate, and the oxygen content of the blood.

General anesthesia can be given as an inhaled gas or as an injected liquid. There are several drugs and gases that can be used alone or in combination. The potency of an anesthetic is measured as minimum alveolar concentration (MAC). "Alveolar" refers to tiny sacs in the lung through whose walls gas exchange with the bloodstream takes place. MAC is the alveolar partial pressure at which 50 percent of humans will not move from a painful stimulus, like a skin incision.

35. What makes us right-handed or left-handed?

Researchers say it is in the genes. It's the same reason some people are brown-eyed and some are blue-eyed. The Human Genome Project tends to support the theory that a single gene is responsible for handedness. One of every ten people is left-handed, with males outnumbering females 1.2 to 1. Being left-handed implies a preference for using the left hand for writing, pointing, throwing, and catching. Left-handers use the left foot for kicking, the start of running, walking, and bicycling. Being left-handed also means having a dominant right side of the brain. Left-handers may also have a dominant left eye, which they use for camera sights, telescopes, and microscopes.

In the past, there was not much sympathy for lefties. Sometimes there was enormous effort by parents and teachers to force left-handed kids to write right-handed, which led to rebellion, frustration, stuttering, dyslexia, and hatred of school. Kids were often labeled clumsy and awkward.

We seem to be much more enlightened about left-handedness these days. Most parents seem to let their children find their own handedness and accept the outcome. Left-handers even have their own day, August 13, on which they can celebrate their uniqueness.

Still, most tools, utensils, and office gear are made for right-handed people. When pants have only one back pocket, it's always on the right side. Lefties have to reach for it with the "wrong" hand. Piano keys are arranged so the right hand plays the melody and the left hand plays the accompaniment. Cars have the stick shift on the right. Camera shutter buttons are on the right. The most frequently used keys on a computer, such as enter, backspace, arrows, and the number keypad, are all on the right.

"Southpaw" has its origins in 1880s baseball slang, when baseball diamonds were often arranged so the batters would face east, to avoid looking into the afternoon Sun. The pitcher's left hand, or "paw," would therefore be on the southern side.

Finally, superior creativity, genius, and career success have been associated with left-handedness. There are long lists of famous left-handed people, especially in the arts and sciences. They include Julius Caesar, Michelangelo, Albert Einstein, Ted Williams, Ronald Reagan, and Jay Leno. One of my three brothers is left-handed, and he makes more money than I do! My four-year-old grandson is left-handed, and his grandparents know that he is a genius! But you could make endless lists of right-handed success stories, so I'm not convinced that there is any real advantage or disadvantage of being a lefty.

36. Why does our hair go gray or white when we get old?

Some people get gray (or white) hairs in their twenties, and others still have dark hair well into their seventies and eighties. The process of graying can occur gradually over many years as individual hair follicles stop producing color, or it can happen within a matter of months or a few years. How long depends a lot on genetics. Also, our changing hair color depends on how dark the rest of the hair is to begin with.

Individual hairs don't actually "turn" gray—they grow in that way. Every day, hairs fall out and new ones emerge in their place. As the hair grows in the follicle, color is deposited into the new growth in the form of two substances, melanin and pheomelanin, which all people have in varying quantities. Melanin produces the hair shades blond, brown, and black, depending on the concentration of pigment in your hair. Pheomelanin produces red hair and the reddish undertones seen in hair. When one of your follicles stops producing these colored pigments (usually with age), the next hair growing out of that follicle will grow in gray. Why this happens is one of the mysteries of aging.

Having gray hair doesn't mean a person is not healthy. You may choose to color your hair for cosmetic reasons, but gray hair is nothing to be alarmed about.

A number of external causes can make hair gray. According to most sources, smoking is a big reason. People who smoke are four times more likely to be prematurely gray. (Some evidence, by the way, links smoking to early baldness, too.) Early graying (or balding, for that matter) is not a sign of early aging. Despite some reports, no solid link has been found. Most people, especially those with dark hair, will begin to notice a few white hairs by their late twenties or early thirties.

37. How do we see color?

The human eye and brain work together to perceive color. The retina is the light-sensing structure on the back of the eye. The center of the retina contains about six to seven million cones. Different ones are sensitive to red, green, and blue light.

Rods, located on the periphery of the eye, are sensitive to dim light and handle vision in low light. Each eye contains about 130 million rod cells. When light hits the rods and cones, complex chemical changes create electrical impulses that are sent to the brain via the optic nerve.

White light—the visible part of the light from sources such as sunlight, incandescent light, and fluorescent light—is made up of seven colors. They are the ROY G BIV colors of red, orange, yellow, green, blue, indigo, and violet. Waves that are longer than red, called infrared, are not detected by the human eye. Waves shorter than violet, named ultraviolet, also cannot be seen by the human eye.

The ROY G BIV colors were named by the English physicist Isaac Newton. Of those seven colors, three are considered the additive primary colors for light. They are red, green, and blue. These same three colors are used in color television and color display panels

In about 1680, Newton found that color is not "built into" an object. The red is not "in" the apple. Rather, the surface of the apple is reflecting some wavelengths of light and reflecting other wavelengths of light to the eye (see page 77). The red apple reflects red light to the eye and absorbs the remaining six colors. We perceive only the reflected light. So we say the apple is red. An object appears white if it reflects all the wavelengths (colors). An object is black if it absorbs all the colors of the visible spectrum. A yellow object will reflect red and green and absorb blue.

The subtractive primary colors for printing, paints, pigments, and dyes are cyan, magenta, and yellow. These are the inks used on rollers when color pictures in magazines and newspapers are printed. These colors are often referred to as CMYK. The K is for black ink.

38. **Will we ever find a cure for cancer?**

All of us are touched by cancer, either personally or through relatives and friends. Cancer is the second leading killer in the United States, right behind heart disease. One of the problems with a cure is that there is no single type of cancer. Cancer is really an assortment of about two hundred diseases. But they all have one thing in common: abnormal and uncontrolled growth of cells. Cancer is cell division gone wild! The growths and tumors destroy body tissue and may spread to other parts of the body.

A large study of almost ninety thousand twins was conducted in Norway, Sweden, and Denmark in order to determine the degree to which genetics play a role in cancer. Twins are ideal subjects for study because they have similar—and in the case of identical twins, nearly identical—genetic makeups. The researchers found that genes don't play as big a role as once thought. It's environmental factors that a pair of twins don't share that are far more important. Smoking, eating too much, eating the wrong foods, lack of exercise, and exposure to pollution and radiation were far more significant factors than genetics. The twins who generally took better care of themselves had a 90 percent chance of not getting the same cancer as their siblings.

The study suggests that we have some control over whether we get cancer and that we can reduce the risk. Early detection plays a large role in the cure rate for cancers. In many cases, the air we breathe and the water we drink are much better than they were just thirty or forty years ago. But there's a long way to go. We continue to ingest a lot of junk chemicals in the food we eat. Just take a look at the content labels on, say, crackers, ice cream, soda pop, or meats. Our diets are loaded with way too much fat. Our cars, trucks, and coal-fired power plants spew tons of chemicals in the air. Our cleaning supplies, plastics, and many other household items contain dozens of harsh chemicals.

Besides early detection, there seem to be two approaches to fighting cancer. First, there's the vaccine approach, which aims to boost the patient's immune system by triggering the production of special cells that kill cancer cells and prevent relapses. The United States Food and Drug Administration has approved two types of vaccines to prevent cancer: vaccines against the hepatitis B virus, which can cause liver cancer, and vaccines against the virus that causes cervical cancer. Will we find a universal vaccine that will do the same for cancer as vaccines did for polio and smallpox?

The second approach aims at cutting the blood supply to tumors. Drugs like Avastin inhibit the formation of new blood vessels that feed tumors. This is a simpler method than the vaccine idea, which involves a lot of very complex biological processes that are not thoroughly understood.

I queried several doctors about the outlook for cancer treatment, and they all said basically the same thing. They think we're making great progress in treating all types of cancers, some more than others. They speculate we'll have a cure for most cancers in twenty to fifty years. We'll have techniques and devices that will spot even a few cancer cells, and we'll have the tools to go after those cells. The hardest cancers will be the kinds that, like AIDS, compromise or attack the immune system. And even present treatments using surgery, radiation, and chemotherapy are getting better.

39. How tall can people grow?

The tallest person on record is Robert Wadlow from Illinois, who died in 1940 at the age of twenty-two. He stood eight feet, eleven inches tall. At the present time, there is a thirty-three-year-old man living in the Ukraine who is eight feet, five and a half inches tall. But he grew one foot fairly recently, so he might surpass Wadlow. Both men owe their extreme height to a condition called acromegalic gigantism. Height is determined by genes, hormones, and nutrition. We humans stop getting much taller soon after puberty. Growth of bones is limited by sex hormones, which tell the ends of bones to stop growing. The average age we are at this stage differs across genders: For boys, it is about age eighteen, and for girls, it is about age sixteen. Both Wadlow and the Ukrainian, Leonid Stadnyk, had a tumor on the pituitary gland. The tumor destroys cells in the pituitary gland that stimulate the release of sex hormones. So the bones never get the signal to stop growing.

Marfan syndrome is a genetic disorder of the connective tissue. People with Marfan syndrome tend to be tall, with long limbs and long fingers. Abraham Lincoln is believed to have suffered from Marfan syndrome.

Medical experts figure that it would be hard for a nine-foot person to live very long. There would be very high blood pressure in the legs, which would lead to burst blood vessels and varicose ulcers. It was an infected ulcer, for example, that killed Wadlow.

Even with modern antibiotics to control infections, an extreme height puts a big strain on the heart. It's a huge job for the heart to pump blood up a height of seven or eight feet.

Some teens go through growth spurts. This growth starts on the outside of the body and works in. Hands and feet are the first to expand; needing new shoes is a sign of a growth spurt. Next, arms and legs grow longer. Then the spine lengthens. Finally, the chest and shoulders widen on boys, the hips and pelvis on girls.

Can a concerned young person increase their height? Yes, to some small extent. A diet rich in fruits and vegetables, cereals, and meat is helpful. Plenty of sleep is crucial. Lack of sleep limits growth hormones. And there are some specific stretching exercises that one can do.

Can human growth hormones (HGH) make a person taller? Yes, if taken during the early years, when a person is growing. But it is tricky, and there are a lot of fraudulent promoters out there. HGH should be prescribed only by a medical specialist.

There are advantages to being tall. Short people are at a disadvantage when seeking jobs. Women are said to be generally more attracted to taller men.

Of the twenty-eight presidential elections from 1900 to 2011, the taller of the two candidates has won eighteen times, the shorter candidate has been elected eight times, and two have been the same height.

40. Sometimes when I fall asleep, I awake with a muscle spasm. What causes this?

I think this has probably happened to most of us at one time or another. The most common type of muscle spasm seems to be leg cramps, which are, like all muscle cramps, contractions of the muscles. According to several reliable medical websites, the usual causes are dehydration, use of diuretics, overuse of muscles from heavy exercise, muscle fatigue, stress or anxiety, and lack of vitamin C. Some low-carb diets, like the Atkins diet plan, lack vitamin C. Drinking too much coffee, a diuretic, can be a big causal factor.

Recommendations for relief include stretching and massaging the leg muscle to help the muscle relax. Putting weight on the "cramped" leg and bending the knee may help. A cold pack relaxes tense muscles (but keep a cloth between it and your skin). Later, apply a warm towel or heating pad if there is pain or tenderness.

Restless legs syndrome is a neurological disorder characterized by an irresistible urge to move one's body to stop uncomfortable or odd sensations. It most commonly affects the legs. Movement is the most common activity that brings relief.

A good website for basic concerns about health issues is www.my.webmd.com. It suggests seeing a doctor if muscle cramps or spasms persist.

41. How fast can a person run?

The title "fastest man alive" did belong to Jamaican sprinter Asafa Powell. Powell was clocked at 9.77 seconds in the hundred-meter dash on June 14, 2005, at the Olympic Stadium in Athens. The Athens track is said to be one of the fastest in the world. In May 2008, Powell did the hundred meters in 9.74 seconds. If my calculations are correct, that comes out to about 23 mph.

Today the title of "fastest man in the world" is assigned to Jamaican Usain Bolt, who ran the hundred-meter dash in a time of 9.58 seconds in Berlin in 2009.

The American Justin Gatlin won gold at the 2004 Summer Olympics in Athens with a time of 9.85 seconds in the hundred-meter run, currently making him the third fastest man of all time. Gatlin would jump over fire hydrants while growing up in Brooklyn, New York. The amazing University of Tennessee track star broke many short-distance records in the 2005 season.

The "fastest woman alive" was Florence Griffith Joyner, who did the hundred-meter dash in 10.49 seconds in 1988. Joyner died in 1998 of suffocation from an apparent epileptic seizure. She was only thirty-eight years old.

The hundred-meter race is an ideal distance to establish a human's top speed. Any longer distance would have the sprinter not running their maximum speed for the entire race. Any shorter distance would mean that a considerable part of the run is used just to get up to top speed.

These speeds are records for humans moving unassisted. The fastest humans alive are the three astronauts who returned from the moon on the Apollo 10 mission on May 26, 1969. Gene Cernan, Thomas Stafford, and John Young were moving at roughly 11,000 meters per second, or about 25,000 mph, just prior to their entry into the Earth's atmosphere.

42. When you look at something for a long time and then look away, why is the image still in your head but in a different color?

We often call these images afterimages or retinal fatigue images. And they're a lot of fun to look at.

The generally accepted theory for why they appear is the retinal fatigue idea. There are three types of color receptors, or color-sensitive cells, that make up the cones on the retina of the eye. Some of these cells are sensitive to red color, some to green color, and some to blue color. Red, green, and blue are the primary colors when it comes to light (see page 49).

When we stare at any particular color for too long, the receptors, or cells, for those colors get "tired," or fatigued, and do not work very well. We may not notice this until we look at a white background, which reflects light of all colors. The receptors that are tired do not transmit the color they are sensitive to along to our brain, so we only see the other colors, which combine to form the complementary, or inverse, color of the image. For example, if we stare at a green-colored object for thirty seconds or more, then look away at a white screen or white sheet of paper, we see a magenta, or purple-like, color, a combination of red and blue. If we look at something blue for a period of time, then glance at a white page, we see a yellow image. Yellow is the complement, or opposite, of blue.

There are some excellent websites that have afterimages and a good explanation of how they work. This is an excellent one: http://www.colorcube.com/illusions/aftrimge.htm. The most famous of the retinal fatigue images is the reverse American flag. Stare at the green, black, and yellow colors for a time. Then fix your gaze on a white sheet of paper, and the familiar red, white, and blue Old Glory appears.

Doctors and nurses in the hospital operating room wear scrubs that are blue-green, or cyan, in color. The doctor might be looking at red blood for a long time and under bright lights. If the doctor looked up and saw an assistant's white scrub suit, there would be a disturbing blue-green afterimage. But if the assistant's clothes are already blue-green, the afterimage is barely noticeable.

I always considered eyesight, or vision, to be one of God's greatest gifts. Then God threw in color for extra credit!

43. Why do people's past injuries hurt when a storm is coming?

Because my field of expertise is the physical sciences, I like to consult experts when it comes to questions in the biological sciences. Tim Kortbein, a physical therapist I know at our local hospital, says that there is a definite relationship between weather changes and people's complaints about their joints and muscles.

The barometric pressure drops as a storm or bad weather is approaching. Joints may not be able to adjust to the pressure change, and then the soft tissue and fluid around joints expand, irritating nerves and causing pain. This is especially true of any joints that are arthritic. Also, the metal that is in knee, ankle, or hip replacements can cause pain if the weather turns cold. This is particularly true for patients who have had recent implants or replacements. For the first few years, the bone is adapting and growing around any metal prosthesis. Bone activity is sensitive to weather and pressure changes.

If an injury is not rehabilitated properly, problems may show up later in life. An injury to a leg or ankle may cause a person to favor that leg. A change in walking style may then result in overuse and impairment. Muscles work in pairs. For example, the biceps in our upper arms contract while the triceps relax. If an injured muscle is not properly rehabilitated, a stronger muscle may overpower the weaker one it is paired with.

44. Why can't we keep our eyes open when we sneeze?

Sneezing is a reflex response that expels an irritant from the nose (see page 18). It involves numerous muscles in the face, throat, and chest. The sneeze reflex uses muscles that close the eyes as well. A reflex action is a series of reactions that are programmed in the body by the brain. This makes it nearly impossible to voluntarily keep our eyes open when we sneeze. However, for some people, it is possible to ward off a sneeze by pinching the end of their nose.

The speed of a sneeze was estimated to be, on average, 100 mph, or about the same as the best fastball in baseball. However, MythBusters Adam Savage and Jamie Hyneman conducted a scientific probe of the venerable sneeze and found speeds of 35 mph for Adam and 39 mph for Jamie. That is far less than the previously accepted 100 mph. In addition, the sneeze traveled seventeen feet for Adam and thirteen feet for Jamie, considerably less than what conventional wisdom would expect.

Thomas Edison came up with the idea of movies from watching someone sneeze in 1888. He was studying the successful motion-sequence of still photographic experiments of Eadweard Muybridge and Etienne-Jules Marey. In 1889, Edison was viewing sequential pictures and guessed that if you viewed them rapidly in sequence, you might be able to make a moving picture.

Edison assigned the task to one of his most capable assistants, a young Englishman by the name of W. K. L. Dickson. In January 1894, Dickson produced a short film entitled *The Edison Kinetoscopic Record of a Sneeze.* The person doing the sneezing was another Edison employee, Fred Ott. The film piece was then deposited in the Library of Congress and is the earliest surviving copyrighted motion picture.

45. Are tanning booths safe?

The short answer is: no! Tanning is a response to injury and is nature's way of protecting us, so tanning booths work by causing injury to the skin. Tanning occurs when enzymes stimulate cells in the skin to make melanin, a dark pigment that absorbs ultraviolet (UV) light (see page 34). The job of melanin is to protect the DNA in the lower layers of skin cells. Melanin is a UV absorbent and acts as an antioxidant.

UV light from the Sun or a tanning booth penetrates the upper layers of the skin and damages the DNA inside skin cells. Repair enzymes are sent to get rid of the injured DNA and aid in making new DNA. Sometimes, because of the damage, these repair efforts go wrong and create cancer.

There are some additional drawbacks to tanning, whether in tanning booths or under the Sun. Chronic exposure causes a change in the skin's texture by damaging the connective tissue, causing the skin to become leathery and wrinkled. Tanning can bring on age spots, and it can damage the body's immune system. Exposure in tanning booths can even produce cataracts.

Experts advise that we don't spend any time in tanning booths—which the World Health Organization calls a "known human carcinogen"—but some sunlight is good for us, even necessary. Sunlight stimulates the production of vitamin D; tanning booths, on the other hand, do little for our vitamin D levels, because they expose us mainly to UVA light, a type of ultraviolet light, and it's UVB that stimulates vitamin D. Lack of vitamin D causes rickets in children, which produces soft and malformed bones.

46. What is the doctor or nurse listening for through a stethoscope while taking your blood pressure?

The clinician pumps up the blood pressure cuff with air using a hollow squeeze bulb that allows the air to flow in only one direction. A tiny check valve prevents air from coming back into the bulb. The inflated cuff cuts off the flow of blood in the arteries and veins. Then the doctor or nurse slowly reduces the pressure in the cuff. At the point where the blood can barely squeeze through the compressed blood vessels, this flow is turbulent. The chaotic movement of blood through the blood vessels creates noise.

This is what the health practitioner listens for. When they first hear the turbulent noise, they take the higher blood pressure reading (the systolic pressure). When that turbulent noise disappears, it means that the smooth flow of blood has returned in the veins and arteries, and the practitioner notes the lower second reading (the diastolic pressure).

Blood pressure is the pressure, or push, of the blood against the blood vessel walls. It is one of the four vital signs, the other three being body temperature, breathing (respiratory) rate, and pulse rate (heart rate).

The common device used to take blood pressure is known by a massive six-syllable word: sphygmomanometer. The numbers used for blood pressure originally represented how high, in millimeters, the pressure exerted by the pumped air could push a column of mercury (in a tube of a standard diameter) upward; the unit of measurement for blood pressure is millimeters of mercury, or mm Hg. Today, most sphygmomanometers do not use mercury but instead use an aneroid (without liquid) or electronic device.

Blood flow in our body is somewhat like the flow of water in the pipes of our house. There are gauges that measure the pressure of that water against the pipe walls. The "plumbing" of the human circulatory blood system can be over one hundred thousand miles long, depending on the person's size.

For many years, doctors considered a normal blood pressure to be about 120/80, which can be stated as "one-twenty over eighty." The first number, 120, is the systolic reading, which shows the peak pressure in the arteries, when the ventricles are just starting to contract, pushing blood to all parts of the body (see page 26). The 80 number represents the diastolic pressure, the minimum pressure in the arteries, when the ventricles are relaxed and filling up with blood.

In our example above, if a mercury-filled tube was placed against the artery wall, the blood pressure would push that column of mercury up 120 mm, or nearly five inches, when the heart is working hard, and 80 mm, or a bit over three inches, when the heart relaxes.

Various factors affect blood pressure, including age, sex, height, exercise, sleep, diet (particularly sodium), disease, being overweight or obese, drugs, alcohol, smoking, emotional reactions, stress, digestion, and time of day. Most people have a higher reading in the afternoon and a lower one at night. In children, blood pressure varies with height. As we get older, the systolic pressure rises and the diastolic falls some. Many people's blood pressure is sensitive to salt and other sources of sodium in their diet. For these people, sodium in the food they eat tends to raise their blood pressure and potassium helps lower it. You can see how much sodium is in a serving of a packaged food by looking at the Nutrition Facts label.

Personally, two things can really spike my blood pressure upward. The first is filling in Line 76 on the standard IRS 1040 form, "Amount you owe." The second is seeing, in my rearview mirror, those alternating flashing red and blue lights on a car tailing very close behind.

47. Why does liquid come out of some people's noses when they laugh?

There is a saying that goes: "The joke was so funny it caused him to have nose cola." Excessive jocularity can cause water, milk, or soda to come out the nose.

Food and drink should go down the esophagus and into the stomach. Air should go down the trachea, then into the branching bronchial tubes, and end up in the lungs. But those two tubes, the trachea and the esophagus, are quite close to each other. So the drink can go down the trachea and bronchial tubes—we say it goes "down the wrong pipe"—and then be expelled out the nose. Instead of being swallowed, the fluid goes up the nasal passages and out through the nostrils.

There is no real danger medically. Perhaps there is some social embarrassment for an adult. Absolutely no embarrassment for a preteen or teen boy!

The human body has safety devices that prevent food and drink going down the wrong way. There is a flap in the back of the throat called the epiglottis. The epiglottis covers the larynx when you swallow food. Sometimes, if a person is talking or laughing while swallowing food or drink, the epiglottis does not block the larynx completely and food enters the wrong pipe. It can spurt out the nose.

The larynx sits atop the trachea. The larynx contains the vocal cords that we need to speak and sing. These vocal cords close up and go into a spasm if food or fluid gets to them. As a final defense, if food or drink gets into the trachea or windpipe, you have a cough reflex that should

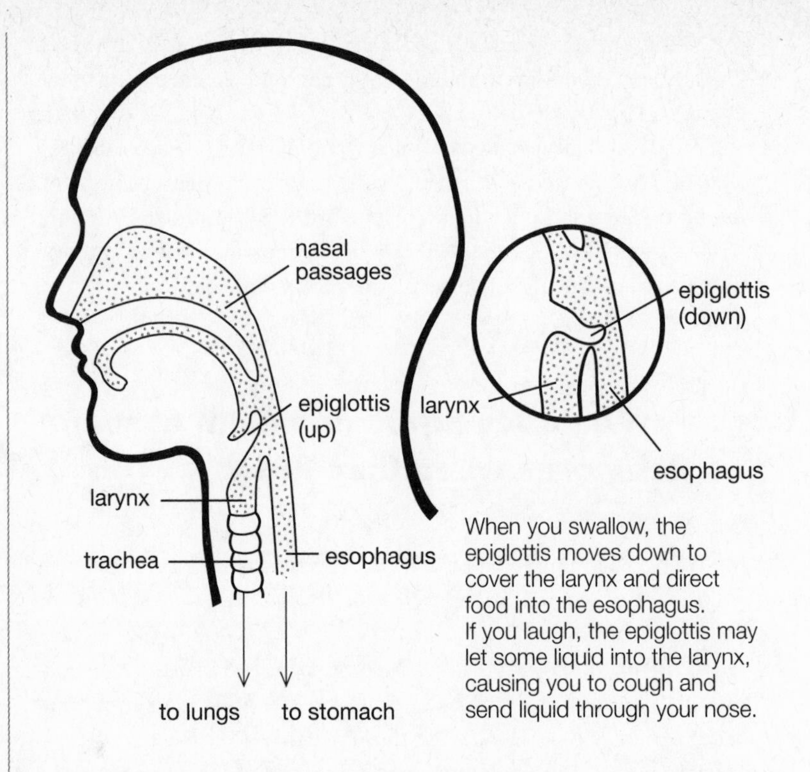

When you swallow, the epiglottis moves down to cover the larynx and direct food into the esophagus. If you laugh, the epiglottis may let some liquid into the larynx, causing you to cough and send liquid through your nose.

expel any foreign matter. These three mechanisms can be dismantled by excessive alcohol, which will allow food and drink to get into the bronchial passages and cause death by asphyxiation.

What about a cow? If a cow laughed really hard, would milk come out her nose? The answer is no. It's udderly impossible.

48. How many muscles are in the human body?

Answers vary from 640 to 850, depending on how you count them. Muscles make up the engine of the body by turning energy into motion. It is impossible to do almost anything without them: talking, eating, writing, running, and dancing are all done with muscles. In addition, we have muscles that act involuntarily for breathing and for keeping our heart beating.

Muscles are divided into three body systems. Skeletal muscles move the bones and the facial muscles. These muscles have striped, or striated, fibers and are called voluntary muscles because the human brain consciously controls them. These muscles are the ones that get injured in physical activity or sporting events. They make up about 47 percent of body mass in men and about 38 percent in women.

Smooth muscles are found in the stomach and intestinal walls, in vein and artery walls, and in other internal organs. These are involuntary muscles; we do not generally control them, so we don't need to consciously think about them. Instead, they get their signals from the autonomic nervous system, which tells muscles like the secretory glands to pump hormones directly into the bloodstream and the digestive smooth muscles to push food through the intestines. These muscles are also located in the urethra, bladder, esophagus, and bronchi.

Cardiac muscles, or heart muscles, contain both striped and smooth muscle tissue, and they consist of a branched, continuous network of muscle fibers, with prominent bands crossing the fibers at various intervals. Purkinje fibers form the impulse-conducting system of the heart, and they're totally involuntary, thank goodness. No need to "tell" the ticker to keep ticking.

Muscles come in different sizes, and some muscles work harder than others. The busiest muscles in the body are the ones that do the blinking for us. We blink 15,000 to 22,000 times every day, both voluntarily and involuntarily. The smallest muscle is the stapedius muscle, in the middle ear. The largest is the gluteus maximus, or buttock.

If we work our muscles too hard, the cells run out of oxygen. This starts a fermentation process that produces lactic acid. The buildup of lactic acid in the muscles causes soreness and stiffness.

Muscle exercise also affects bone growth. Ultrasound studies on major league pitchers indicate that their pitching arm's bone is larger and stronger than their non-throwing arm's by as much as 30 percent.

Muscle fibers are long cylinders compared to other cells in the body. Their job is to create contraction, and, when necessary, to relax. All muscle activity is electrical; the brain sends an electric signal to the muscle and tells it to contract. A little or a lot, depending on the nature of the signal.

Multiple sclerosis, or MS, is a disease in which these signals traveling along the nerve fibers are interrupted or distorted by scar tissue. Muscular dystrophy, or MD, is a loss of muscle mass. The muscle wastes away, getting weaker and weaker. MD is an inherited condition, passed down through families.

49. Why do we blink?

Blinking is a way of providing moisture to the eye. The average blink rate is ten to twenty times per minute. Blinking is involuntary, not something we consciously do. The blink lasts a fraction of a second; we rarely even realize it's happening, and it is fast enough that it does not interfere with our vision. We all blink so often that we tend to ignore it altogether; we hardly ever notice blinking in other people.

Blinking is different from batting one's eyes. Batting is opening and closing the eyes several times in rapid succession, usually deliberately.

During a blink, the upper eyelid comes down over the eyeball and creates suction across the eye to prevent it from drying out. Moisture, provided by tear glands through tiny ducts, flows over the surface of the eye. The eyelids have about twenty-five oil-producing glands, located between our eyelashes and not visible to the human eye, and this oil mixes with the water from the tear glands. So blinking provides water and oils for the eye, similar to a farmer irrigating crops.

Blinking also affords protection for our eyes. Blinking shields the eye from dust and debris, and our eyelids function as windshield wipers when a particle does come in contact with the eye. But the first line of defense is eyelashes. Those curved hairs are dust catchers that prevent dust in the air from hitting our eyeball. Eyelashes are common in the animal kingdom. You might have noticed that cows and horses have long eyelashes, and camels in the desert have extremely long eyelashes to protect their eyes from sand.

There is strong evidence that blinking is useful when we change focus. We may be looking into the distance, say, out the window. Then we shift our attention to something close, say, a computer keyboard. Blinking helps the eye to settle into a new point of focus. People have been observed to blink more often during changes in focus of their eyes.

From a research group in Japan comes evidence published in the *The Proceedings of the National Academy of Sciences* that blinking may help us gather our thoughts and focus our attention. They suggest that blinking provides something like a mental resting place that shuts off visual stimulation and refocuses our attention.

Rapid blinking may indicate a sign of nervousness in a person. There is some evidence that a person exhibiting rapid blinking may be hiding the truth or telling a lie.

There are some recognized disorders associated with excessive blinking. The cause could be the side effects of medication or some psychological unease. Other causes are fatigue, allergies, eyelid inflammation, Tourette's syndrome and other tic disorders, brain tumors, and seizures. Chronic dry eyes can also bring on excessive blinking. Being outdoors in windy conditions can dry out the eyes in a hurry. This can occur when the water in tears evaporates faster than the oil layer in the tears. Advanced age, contact lens use, and certain medications are other common causes of dry eyes. People might find temporary relief by using eyedrops or artificial tears.

50. Why do people turn red when they get embarrassed?

We've all been there: feeling stressed, scared, nervous, ashamed, or embarrassed. Any strong emotion can induce blushing. That includes hearing a flattering remark, making a speech, being the center of attention, or feeling self-conscious. Getting angry can bring on a blush, too. In every case, the face turns various shades of crimson.

Science doesn't have all the answers about blushing. Medical people say that blushing is a result of both the "fight or flight" response and social behavior. "Fight or flight" is the theory that animals react to threats with a reaction of the nervous system that prepares the animal for fighting or fleeing.

Blushing in humans is triggered by social situations, not threats to health or life. Animals don't blush. So even though some responses may be inherited traits from our distant past, blushing in humans is related to something animals don't have, namely, a moral consciousness.

The body secretes more adrenaline as part of the sympathetic nervous system response. This system is involuntary; you don't have to think about it. The pupils of the eye grow bigger, allowing you to take in more visual information. The capillaries that carry blood to the skin widen. There is increased blood flow to the neck and limbs. The heart rate increases, and breathing becomes faster. The blood vessels in the face dilate, opening to more blood flow, and the face reddens.

Blushing is more noticeable in people who have fair skin. Northern Europeans—Nordic, English, and German people—generally fall in this category.

There can be other reasons for a red face. People who suffer from hypertension may show a reddened complexion. And as the body heats up during exercises, it tries to cool down both by sweating and by sending more blood to the surface of the skin. Alcohol can cause arteries to widen, too. Some Asian populations don't have the ability to break down alcohol in the liver. The toxic material enters the bloodstream and causes blushing.

51. **How am I the only one in my family with blond hair?**

Blond hair is common in infants and children. Babies are often born with blond hair even though their parents have dark hair. That blond hair turns darker with age so that blond-haired kids have dark or even black hair when they are in their teen years.

The chemical melanin gives hair its color. The more melanin, the darker the hair color (see page 34).

Hair color historically was tied to geographic areas. Hair color in southern Europe, North Africa, the Middle East, and the Americas was dark brown or black. In central Europe, lighter brown hair was common. Blond or yellow hair, and even red, was prevalent in northern Europe, Britain, and Ireland. The country of Lithuania has the highest percentage of people with blond hair. To this day, blond, or "fair," hair is common in northern Europe: Norway, Finland, Sweden, Denmark, and the Netherlands. Very pale hair is known as "Nordic blond." Blond hair in northern Europeans is also associated with eye color. Fair-haired people tend to have blue, green, or light brown eyes.

The Human Genome Project, completed in about 2003, sought to identify the approximately 20,000–25,000 genes in human DNA. The Project also determined the sequence of the three billion pairs of chemical bases that make up our DNA. This project turned up very strong evidence that two gene pairs and a few other specific chromosome regions control human hair color.

An allele is one member of a pair of genes that a person has. Each can be dominant or recessive. One gene pair is a dominant brown and a recessive blond. A brown-haired person can have either two brown alleles or one brown allele and one blond allele. This model explains why two brown-haired parents can have a blond child: Both parents have one of each allele, and the child received two recessive alleles, one from each parent. The other gene pair is a non-red/red pair.

But this still doesn't explain why hair color changes as kids grow older. If genes can't explain this, there may be some environmental factors involved. One such factor is strong sunlight, which tends to lighten hair as it washes the color out. Some kids who run around outside in the summer will have pale blondish hair by the end of summer, but the pigment returns as new hair grows in the wintertime. Their hair is back to its brown color by Christmas.

There's still one more mystery: the spelling. While some people use "blond" and others use "blonde," both are correct, although there's a tendency to associate "blond" with males and "blonde" with females.

52. What makes our eyes go bad?

Well, I feel your pain. It happened to me, also. I was the only one in my class to wear glasses during second and third grade. Of course, there were only four kids in my class in that one-room country school in Wisconsin. After I had been wearing glasses for about eighteen months, a real stroke of good luck came to pass. I fell down on the ice at school, at the little frozen pond created when water filled in the home plate area of our softball field and the weather turned cold. It made a nice round skating area about eight feet across. We threw snow on the ice to make it a bit more slippery.

Too slippery for boy Scheckel: Down I went, and I shattered one of the glass lenses. These were real glasses, not the plastic lenses we have these days. After the obligatory scolding, my dad and mom took me to the eye doctor in La Crosse and had my vision retested. The doctor said I didn't need glasses anymore! Perhaps there is a silver lining in home plate ice ponds.

That familiar eye chart was developed during the Civil War by Dutch ophthalmologist Herman Snellen. It's still called a Snellen chart. If you can read the line that says 20 without glasses, you have normal vision, or 20/20 vision. You can read at 20 feet what everybody with normal vision can read at 20 feet. If you have 20/40 vision, you need to be 20 feet from what people with "normal" vision can read at 40 feet. That's not too good.

Test and fighter pilot and sound-barrier breaker Chuck Yeager has 20/10 vision. He can read the letters on the line labeled 10 at a distance of 20 feet, while normal-visioned people need to be 10 feet from the chart. That is top-notch eyesight.

Most kids who need glasses are nearsighted, or myopic. Objects far away look fuzzy. Activities that involve their close-up vision, such as reading a book or looking at a computer screen, are OK. The corneas and lenses in their eyes focus the image in front of the retina, which is that light-sensitive layer of cells on the backs of the eyeballs.

The lenses needed to correct nearsightedness are generally concave, thinner in the middle of the lenses and thicker at the edges. It's just the opposite for a farsighted, or hyperopic, person. The cornea and lens do not bend, or refract, the incoming light rays sufficiently and tend to focus them behind the retina. Farsighted people can see things clearly far away, but their close-up vision is not good. They typically need a convex lens, one that is thicker in the middle and thinner on the outside edges.

Astigmatism is another vision defect that might cause a person to wear glasses. The front surface of the eye, the cornea, is somewhat oval instead of being round. It bends entering light rays different amounts depending on the direction they come in from. A person with astigmatism has distorted vision. Lenses can be ground to correct for the astigmatic error. Astigmatism can also be corrected by laser surgery. The cornea is reshaped by removing bits of the cornea using a highly focused laser beam.

Two types of laser surgery are used to correct for refractive errors, including nearsightedness, farsightedness, and astigmatism: photoreflective keratectomy (PRK) and laser in situ keratomileusis (LASIK). PRK removes tissue from both the surface and inner layers of the cornea. LASIK surgery removes tissue only from an inner layer. In LASIK surgery, a section of the cornea surface is cut and folded back to expose the inner tissue. A laser removes the correct amount of tissue needed and the flap is placed back into position to heal.

There is one thing that all kids, and grown-ups, too, can do to protect their eyes: Wear UV-protection sunglasses when outside for any long period of time. Studies suggest you'll greatly reduce your chances of getting cataracts later in life.

53. Why does helium make your voice go really high?

Sound is produced when air is moved by our vibrating vocal cords. The pitch, tone, or frequency (all the same thing) of our voice depends on several factors. A person with a bass voice has long vocal cords and lower pitch. One with a soprano voice has short vocal cords and higher pitch. Another factor is the configuration of the anatomy involved in producing our voices. Yet another factor is the gas that we breathe, normally a mixture of oxygen and nitrogen. So the way we sound is determined by a host of factors: the size and shape of our larynx, trachea, mouth, and nasal passages; the length of our vocal cords; and the density of the gas we breathe.

Helium has one-seventh the density of air. So sound travels much faster through helium than it does through air. As a matter of fact, sound moves about 1,100 feet per second through air but about 3,000 feet per second through helium.

The vibration rate of the vocal cords does not depend on the kind of gas that surrounds them. Helium does not make the vocal cords longer or shorter, nor does helium change their tension or stretch. It is the less-dense helium, which serves as the medium for the sound waves, flowing through the larynx that produces this Donald Duck effect. It changes the resonant frequency of the vocal tract, causing a faster vibration and a higher pitch.

Breathing helium from a balloon and producing the cartoon character Donald Duck sound has been going on since people started putting helium in birthday and party balloons. Just a word of caution: In 1996, a healthy thirteen-year-old boy inhaled helium gas directly from a pressurized tank. He just about died from a cerebral gas embolism when bubbles got into his bloodstream and traveled to his brain. Doctors got him into a hyperbaric chamber, the kind that is used by underwater divers who get the bends. He was lucky and made a complete recovery.

In February 2012, a fourteen-year-old eighth-grade girl in Oregon died from inhaling helium from a pressurized canister. The pressure in the tank sends helium with such force that the lungs are ruptured. So don't put your mouth on the nozzle of a helium tank! Also, those helium tanks might have tiny particles of metal in them. And if those get in your lungs, they will stay there. Talk about your heavy metal!

Another gas, sulfur hexafluoride (SF_6), is about five times denser than air and can make a person's voice sound much lower, like the actor James Earl Jones's voice.

54. Why are the pupils of our eyes black?

Near the end of his life, Albert Einstein was asked for the secret of his success. He replied, "I never tired of asking simple questions. I ask them still." At first glance, asking why our pupils are black looks like a simple question. But the answer can be quite complex.

The pupil is the opening in the center of the iris where light enters the lens, which then focuses it on the retina. No matter what color eyes we have, everyone has black pupils. The pupil is black because all the light that enters through the pupil is absorbed inside the eye and none of it is reflected.

Imagine the pupil to be colored, let's say red. White light (sunlight, incandescent light, fluorescent light) consists of the seven ROY G BIV colors: red, orange, yellow, green, blue, indigo, and violet. The red pupil would reflect the red portion of light, which would not enter the eye. This would distort the color of everything we see. It would be like perpetually looking at the world through stained-glass windows. So, as it turns out, the color of the pupil *must* be black in order for us to see the spectrum of colors from red to violet.

Under normal lighting conditions, the diameter of the pupil is about four millimeters. (A millimeter is about the thickness of a dime.) The size can change from about three millimeters in bright light to about eight in dim light. The size of the pupil is responsible for depth of field, sometimes termed depth of focus, or the range of distances over which we can see objects in good focus. A small opening, or aperture, increases depth of field. That's why we squint when we are having difficulty seeing something. In the language of camera buffs, we are "stopping down our eyes."

55. Do identical twins have the same fingerprints?

A lot of people wonder about this. The answer is no. Our DNA contains the instructions for making us who we are. Identical twins have DNA that is almost indistinguishable, because they are formed from a single fertilized egg that splits into two after conception. But fingerprints are not the result of genetics alone. Identical twins have the same genetic makeup (genotype) and are very much alike, but there are subtle differences—enough that people, especially parents, can tell them apart. That's because such characteristics as height, weight, body form, reflexes, metabolism, and behavior (phenotype) are determined by a person's individual genes and by the interaction with nature. This is also the case with twins' fingerprints.

It's that age-old "nature versus nurture" question. How much of what we are as humans is the result of our genetic makeup (nature), and how much is determined by our interaction with the environment (nurture)?

By environment, we're talking about how you're raised, your home situation, what you eat, how you sleep, your siblings, and the air you breathe—in short, everything and everyone around you.

Here's where it gets really interesting. Fingerprints are one of those traits that are the result of development of the baby during pregnancy. Those factors include fetal blood pressure, nutrition, position in the womb, and how fast the fingers are growing by the end of the first trimester. The creation of the patterns of the fingerprint is caused by stresses in a sandwiched sheet of skin called the basal layer. The basal layer grows faster than surrounding layers, and as it does so, it buckles and folds in several directions, forming complex shapes. It's a very random process.

The fingerprints of both identical twins are quite similar, but there are differences in the patterns of arches, whorls, and loops. These differences are caused by the random stresses in the womb—even the length of the umbilical cord has an influence. There are also differences between the fingers on any individual's hand. Probably the most celebrated case was that of the Dionne quintuplets, born in Ontario, Canada, in 1934. The five identical (same DNA) girls all had different fingerprints and handprints.

It is worth adding that fraternal twins develop from two different eggs. Fraternal twins are no more closely related than ordinary siblings. They just happen to share the same growing space for nine months.

56. Why do they give iodine tablets to people who have been exposed to radiation?

Iodine tablets—in the form of potassium iodide (KI)—protect the thyroid from cancer. The thyroid is the body organ most at risk from excessive amounts of radiation. The thyroid is a butterfly-shaped endocrine gland found in the neck near the Adam's apple, lying against and around the larynx and trachea. It regulates metabolism, body temperature, physical growth rate, and brain development.

The thyroid uses iodine to make thyroid hormones, but it can't distinguish between regular iodine and radioactive iodine, the isotope iodine-131 (I-131). The concept behind taking iodine tablets is to fill up the thyroid gland's receptors with regular, good iodine from the tablets, and then they can't take in much of the bad, radioactive iodine. Children and unborn babies are the most vulnerable, because the cells in their bodies are dividing much faster than those in adults, and people whose iodine levels are low are more likely to get thyroid cancer if they are exposed to radiation.

There are side effects from potassium iodide that can be serious, including damage to the salivary glands, allergic reactions, and really bad stomach upsets. Also, iodine pills do not prevent other cancers or protect other organs in the body.

Iodine pills have to be taken before or immediately after exposure to be effective. After the 1986 Chernobyl nuclear accident, authorities waited too long, over a week, before issuing iodine pills. They did more harm than good; radioactive iodine got locked in people's thyroids. Radioactive iodine isn't all bad, though; it is used to treat thyroid cancer and certain other thyroid disorders. However, patients must then take precautions to avoid exposing the people around them to it.

In some parts of the world, where iodine is lacking, the thyroid gland can become enlarged, giving the person a goiter. A goiter is an enlarged swelling of the thyroid and often appears as a large growth in the neck area. As youngsters in a one-room country school in the hill country of Wisconsin, my classmates and I were given a purple "goiter pill" every Friday. The soil around the Great Lakes region does not contain much iodine.

The Morton Salt Company started putting iodine in salt as early as 1924. These days edible salt is sprayed with potassium iodate for little more than one dollar per ton.

Another well-known case of widespread radiation exposure affected children born in the 1950s and early 1960s, who have elevated levels of strontium-90 (Sr-90) in their bones and teeth. The United States and the Soviet Union were running nuclear bomb tests in that period; together, they tested 422 nuclear bombs in the atmosphere. Much of that radioactive material went high in the atmosphere, circling the Earth several times via the jet stream and finally settling down in fields. Cows ate the grass, kids drank the milk, and the Sr-90 settled in their bones and teeth. Strontium is just below calcium in the periodic table, and Sr-

90 gets deposited in the same places in the body; it can cause bone and bone marrow cancer and leukemia. Thankfully, the 1963 Partial Nuclear Test Ban Treaty put an end to most atmospheric bomb tests.

57. What is an autopsy?

An autopsy is a medical procedure to determine the cause and manner of a person's death or to evaluate disease or injury. It is a medical examination of a dead body by a medical examiner or forensic pathologist.

An autopsy is performed when foul play is suspected and in cases of poisoning or drowning. Homicides, suicides, drug overdoses, and cases of sudden infant death syndrome (SIDS) are all candidates for an autopsy. Autopsies are also performed when a medical condition has not been previously diagnosed, for research purposes, or for cases where cause of death could affect insurance settlements or legal matters.

Most autopsies are performed by a forensic pathologist. The body arrives at the medical examiner's office or hospital in a body bag or a sterile evidence sheet and is refrigerated if the autopsy is not performed immediately. Every step of the autopsy is documented with notes, photographs, and a voice recorder. The external surfaces of the body are examined, and the body is weighed and sometimes X-rayed.

Examination of internal organs starts with a large, deep, Y-shaped cut from shoulder to shoulder meeting at the breastbone and extending all the way down the length of the torso. The examiner then peels back the skin, muscle, and soft tissue and makes two cuts through the bone, one on each side of the rib cage. They can now remove the front of the rib cage, exposing all the internal organs. Next the examiner makes cuts to detach the larynx, bladder, rectum, and vertebral column. They can now remove nearly the entire organ set and examine the organs one at a time. The examiner can weigh and dissect various organs and prepare tissue samples for examination. The brain can be cut free of the spinal cord, removed, and examined separately.

An autopsy takes three to six hours, and throughout the entire process, the medical examiner is looking for signs of trauma or other indications of cause of death. When the autopsy is done, the examiner puts the organs back into the body or has them incinerated, replaces the rib cage, closes and sews up the chest flap, and replaces the top of the skull and sews up the scalp to keep it in place.

An autopsy does not prevent an open-casket funeral if the examiner performs what is called a cosmetic autopsy. Family members can request an autopsy, but they have to pay for it. The results can affect insurance settlements or serve as evidence in a legal case.

According to the office of my local medical examiner, my county, Monroe County, sends an average of seven bodies per year to the state capital, Madison, to have autopsies performed. The cost to the county for each autopsy is about thirteen hundred dollars. The decision to have an autopsy performed is made by the medical examiner, with input from the police department and district attorney's office.

When President Kennedy was assassinated in 1963, federal agents took control of the body and the investigation. A nonforensic pathologist did the examination of the deceased president, which sparked controversy. Some claim the autopsy was botched. The law now mandates that a certified pathologist from the Armed Forces Institute of Pathology do examinations in federal investigations.

2

Wonders of Our Sea and Land

58. Why does it seem that most thunderstorms happen later in the day?

A very good observation. Yes, thunderstorms occur in the late afternoon and early evening in many parts of the United States, including my home in the Midwest. During the morning hours, the air near the surface of the land is humid and cool. As the Sun comes up, it starts to heat up the ground, which then heats the air above it, and any moisture on the ground in the form of dew evaporates into the air.

It takes the hottest part of the day to generate the most unstable air, caused by a lifting action as warm air rises. In the process, the warm air cools, and cool air cannot hold as much moisture as warm air can. So the relative humidity shoots up. Relative humidity is the ratio of moisture in the air to the amount of moisture the air can hold (100 percent humidity). As the air rises and cools, it has less ability to hold moisture.

At the same time, the dew point temperature decreases. The dew point is the temperature at which the air can no longer hold moisture as vapor, so the vapor must condense into liquid water. At a certain altitude, typically about ten thousand feet, the vapor in the air condenses onto dust particles or oxygen molecules and forms a little visible bead. Those little droplets of moisture make up a cloud.

This lowest altitude of condensation is generally where we see the bottom of those fluffy white cumulus clouds, usually in the summer sky. If enough of this lifting and condensing occurs, the cloud may build into a cumulonimbus cloud, or thunderhead, which is the type of cloud that produces thunderstorms. A line of these thunderheads is called a squall line. Every thunderstorm needs moisture, unstable air, and lifting capability to form.

The average thunderstorm is fifteen miles across and lasts thirty minutes. Nearly two thousand thunderstorms are occurring at any moment around the world. Winds in thunderstorms can exceed 100 mph, and lightning kills an average of fifty-three people per year in the United States.

59. Why is snow white, and where does its color go when the snow melts?

Snow is a bunch of ice crystals stuck together. It's a very complex arrangement. To understand why snow is white, we must be familiar with what happens to light when it strikes any material. The color of anything, including snow, depends on how light interacts with it (see page 48).

Visible light consists of a rainbow of colors, the ROY G BIV colors of red, orange, yellow, green, blue, indigo, and violet that were assigned by Isaac Newton. When photons of light strike an object, they may bounce back (reflection), bounce to the sides (scattering), pass right through (transmission), or give up their energy (absorption). Grass is green because it reflects the green light to our eyes and absorbs all the other colors. Red apples reflect red light to our eyes and absorb the wavelengths of all the other colors.

When light goes into snow, it hits all those ice crystals and air pockets and bounces around, and then some of the light comes back out. Snow reflects all the colors; it doesn't absorb, transmit, or scatter any single color or wavelength more than any other. The "color" of all the light wavelengths combined equally is white. So all the colors coming out are the same colors that go in, combining to make white light.

A few years ago, I visited the famous Mendenhall Glacier in Juneau, Alaska. The glacial ice looked bluish. Ice is just very compact snow, without a lot of light-scattering bubbles. Light can penetrate much deeper into ice than into snow. The deeper the light goes, the more the longer wavelengths, toward the red end of the spectrum, get scattered out, and eventually the reds dissipate, leaving only blue colors to be reflected back to us. So the ice takes on a beautiful, eerie blue tone.

The record snowfall for any one year in the United States is 1,140 inches (95 feet) at Mount Baker Ski Area in northwestern Washington, during the 1998–1999 snow season.

Snow is beautiful. It coats everything in a pure white blanket. It helps farmers and is good for the land, because it has a ton of air pockets. Even though the snow itself is cold, the air that they hold in insulates the ground, protecting seedlings and preventing the frost from going too deep. Snow that falls in the mountains later melts and helps fill the depleted reservoirs of the American West.

You may have noticed how quiet it is outside after a fresh snowfall. In addition to making snow fluffy those air pockets absorb sound, just like the ceiling tile in my classroom. After a few days, sound travel returns to its normal pattern. Many of the fluffed-up ice crystals melt and compact somewhat, so the tiny pockets that absorbed sound are gone.

When I was a kid on the farm, my dad planted fields of oats in April. One year the oats were up about three inches when we got one of those late-spring snowstorms with four or five inches of snow. I thought all the oats would be dead. Strangely, my dad didn't seem to be concerned. Turned out those were some of the best oats we ever had. I remember him mentioning something about the snow adding nitrogen to the soil, which makes sense because moisture helps plant seedlings fix nitrogen in the soil.

60. Why does the Great Salt Lake have salt?

The Great Salt Lake is quite salty because it is a terminal lake, which means it does not have an outlet. Even though the water in the rivers feeding it doesn't seem salty, they have small amounts of salt dissolved in them, which they constantly bring into the Great Salt Lake. This water has no place to go, so much of it evaporates from the lake surface, leaving the salt to accumulate. The lake's three tributary rivers deposit over a million tons of minerals in the lake each year. The water flowing down from the mountains carries these dissolved mineral salts that have been removed from rocks and soil. The Great Lakes in the Midwest would also be salty if they did not have an outlet. Fortunately, the Great Lakes empty into the Atlantic Ocean by way of the St. Lawrence Seaway.

The Great Salt Lake is the largest salt lake in the Western Hemisphere and is located in northern Utah. Its size fluctuates depending on rainfall and snowfall. Its smallest area was about one thousand square miles in 1963, and its largest size was more than three thousand square miles in 1987.

The shoreline keeps changing. In late summer, the lake is smaller because much water has evaporated away. Snowmelt makes the lake bigger in the spring. So the shores of the Great Salt Lake have remained fairly undeveloped, with widespread wetlands that attract migrating birds.

Because of all the salt, the water in the Great Salt Lake is denser than the human body. You cannot sink in it; you'll float like a cork. Only brine shrimp and some algae have managed to live in the lake; it's too salty to support fish. Mono Lake in California, just east of Yosemite National Park, is another very salty lake. It is a terminal lake in a low spot that has no outlet.

But which body of water is the saltiest? The ocean is about 3.5 percent salt, the Great Salt Lake varies from 5 to 25 percent salt, and Mono Lake is 10 percent salt. Even saltier is the Dead Sea, located between Jordan and Israel. It is 31.5 percent salt, and it also has the lowest elevation on earth, at 1,400 feet below sea level. In comparison, Death Valley in California is 282 feet below sea level. But the prize for "saltiest" goes to the Don Juan Pond in Antartica. It's eight times saltier than the Dead Sea!

61. What is the lowest temperature known in nature?

The lowest possible temperature is termed absolute zero, which is −460°F, the same as −273°C, and also the same as zero Kelvin. Scientists have come within a few hundredths of a degree of absolute zero in the laboratory.

Temperature measures the average kinetic energy of molecules. The greater the motion or vibration of molecules in an object, the hotter it is. So outer space has no temperature, because it is a vacuum. The few particles floating around out there would have a temperature of about three Kelvin, close to absolute zero.

On Earth, temperatures have never gotten anywhere near absolute zero. The lowest temperature ever recorded on Earth was −129°F, which is 184° above absolute zero, on July 21, 1983, at Vostok, a Soviet station in the Antarctic. Keep in mind that July is actually wintertime in the

Southern Hemisphere. The coldest temperature ever recorded in the United States was −80°F, at Prospect Creek Camp, Alaska, on January 23, 1971. Prospect Creek is along the Alaskan oil pipeline and is just north of the Arctic Circle. And the lowest temperature recorded in the contiguous forty-eight states was −70°F, at Rogers Pass in Montana on January 20, 1954.

Many physics classes, such as my own, feature demonstrations using liquid nitrogen. The temperature of liquid nitrogen is −321°F, or −196°C. In one experiment, we would put a constant-volume, hollow, stainless-steel sphere in four different liquids, including liquid nitrogen, and record the pressure of the air inside the sphere. We would then graph pressure versus temperature and use extrapolation to determine the value of absolute zero. We would also make ice cream using liquid nitrogen; we'd study the behavior of materials, ranging from balloons to flowers to a rubber handball, in liquid nitrogen; and we would demonstrate superconductivity by levitating a magnet above a super-cold pellet. This superconducting pellet, the size of a checker on a checkerboard, was made of yttrium, barium, and copper. Cooled to 77K, or −196°C, or −320°F, the pellet wafer became a superconductor, losing all resistance to electrical flow.

Magnetic levitation occurs due to the Meissner effect. A magnetic field surrounds the magnet. This magnetic field induces a current in the pellet. Surrounding the current carrying pellet is a magnetic field. These two magnetic fields, one from the magnet and one from the pellet, oppose each other, creating levitation.

Here are my favorite "How cold was it?" jokes:

- I let my dog out last night, and I had to go out and break him away from a tree.
- We had to chop up the piano for firewood, but we only got two chords.
- The mice were playing hockey in the toilet bowl.
- The polar bears were buying fur coats.

62. Why does the horizon look like it touches the ground?

The world we live in is both beautiful and surprising. And sometimes what we see can mislead us. How can the sky meet the Earth? First of all, let's agree that the horizon is the line at which the sky and Earth appear to meet.

Children will often draw a strip of blue sky and a strip of brown ground with a few rudimentary figures on it, but the rest of the page is white. They do not draw the blue sky touching the ground. Teachers sometimes take young students outside to show them that the sky touches the ground.

So it's not easy to believe what we see. We look straight up and see the sky. But in the distance the sky does appear to meet the Earth. It's all a matter of perspective. Long ago, people on shore would watch ships leave the docks and sail away. The last things that disappeared over the horizon were the tops of the masts, which was a clue that the Earth is round, or spherical.

Nature can fool us in other ways, too. The Moon looks gigantic rising over the horizon but seems much smaller when it's directly overhead. This is a result of the well-known Ponzo illusion (see page 115). The brain interprets the sky as being farther away near the horizon and closer near the zenith, or directly overhead. We notice this on a cloudy day. The overhead clouds may be a few thousand feet away, but near the horizon they might be hundreds of miles away. The sky appears sort of bowl-shaped. The Moon on the horizon is interpreted by the brain as being farther away. Because it is the same apparent size as when it's high up, the brain figures it must be physically larger. Otherwise, the distance would make it look smaller.

And railroad tracks seem to converge and come together when we stand between the rails and look down the tracks. As objects' distances increase, their image size on the retina of the eye decreases, until the farthest objects seem to meet at what is called the vanishing point. Objects seem so small they are no longer visible. So parallel lines, such as railroad tracks, are perceived as coming together until they meet, or converge, at the vanishing point.

There is a formula that tells us how far we can see to the horizon. Another way to put it is that we can calculate how far away the horizon is. The distance to the horizon equals ninety times the square root of the distance, in miles, between your eyes and the ground beneath you. So if you are standing on the beach of the ocean or a big lake and you are six feet tall, 6 divided by 5,280 will give you the distance you are above the sand in miles, and if you take the square root of this number and multiply it by 90, you'll find that the horizon is about 3 miles away. If you go to the top of the Willis Tower (formerly called the Sears Tower), which is about 1,450 feet tall, the horizon will be nearly 50 miles from you.

When I was a kid growing up on a hilltop farm in Crawford County, Wisconsin, I often wondered what it would look like if the sky met the water. The biggest hunk of water I had seen at the time was the Mississippi River, and I could see to the other side. It wasn't until I left home and joined the military at eighteen that I saw the Atlantic Ocean. And, sure enough, the sky touched the water, with nothing in between. It really *could* happen after all.

I have a theory about why the sky is blue and grass is green. If the sky were green, we wouldn't know when to stop mowing (a little science joke)!

63. How do you cut diamonds?

Diamonds measure ten on the Mohs scale of hardness, making them the hardest natural substance on Earth (see page 96). Cutting diamonds is both an art and a science that goes back centuries. Cleaving, the basic diamond-cutting process, is the separation of a rough diamond into separate pieces, to be finished as separate gems. The jeweler places a chisel at a point of weakness in the stone and taps it with a mallet, causing the diamond to split. Any misjudgment by the diamond cutter could lead to losing a valuable gem.

A polishing wheel, called a scaif, was invented in 1456. The diamond is held in a dop, a padded holder that protects the diamond while the jeweler works on it. The polishing wheel is kept lubricated with olive oil and coated with diamond dust. It takes a diamond to cut a diamond. The scaif allows diamond cutters to create symmetrical and even facets, which bring out the sparkle and shine of a diamond.

A diamond saw was developed in the 1900s. Diamond saws are steel blades lubricated with olive oil and edged with diamond dust. Some blades are made of a phosphor bronze alloy. The material lost in cutting and polishing is often over half the weight of the rough diamond.

The value of a diamond depends on the four Cs: cut, clarity, color, and carat. The cut refers to the diamond's geometric proportions: its faces, facets, and finished shape. Clarity refers to the flaws in the diamond. Color can range from milky white to yellow. Carat is a measure of weight and size.

The 3,107-carat Cullinan was the largest diamond ever found. Mined in 1905, it was presented to Edward VII of England. Later, it was cut into nine major stones. The most famous diamond in America is the Hope diamond, on display at the Smithsonian in Washington, DC. It was 112 carats in the year 1668 but has also been recut. A carat is 200 milligrams, or 0.2 grams. There are 454 grams in a pound.

Diamond is pure carbon in an organized, highly compressed form. Carbon is plentiful; our bodies are, for example, 18 percent carbon by weight. Diamonds are formed deep inside the Earth, where extreme pressure and heat turn carbon into diamonds. They are brought closer to the surface of the Earth by volcanic eruptions.

The diamond market is dominated by a single entity, the De Beers cartel in South Africa.

64. Which way does the Earth rotate?

The Earth rotates counterclockwise, as seen from above. "Seen from above" means from a position above the North Pole. The rotation of Earth is vitally important to all aspects of life, because as the Earth spins on its axis, every part of the planet has a chance to face the Sun and receive warmth over an interval that repeats in a relatively short period of time.

We can trace the Earth's rotation back to the way stars and planets are born. A newborn star gathers a disk of dust and gas around itself, and the star's gravity sets that dust and gas spinning. The disk begins as a large mass of gas and molten liquid. The center of the disk becomes the Sun, and the outer rings and clumps of matter cool and condense to take on solid form. Any clumps within the larger mass of spinning dust and gas are going to have their own rotation. These clumps form the planets. As each clump collapses onto itself, its volume decreases and its density increases, causing it to spin faster and faster, a concept of motion known as "conservation of angular momentum." Figure skaters make use of this principle when they bring their arms in closer to their bodies to speed up their spin rate. Because space is a vacuum, no force or friction is there to stop the rotation. The Sun and its planets will just keep on spinning forever. For this rule, we can credit Newton's law of inertia: objects in motion tend to stay in motion.

The Moon's gravity affects Earth's rotation. It causes the waters of the oceans to wash up the shore and back down the shore. This tidal friction slows down the rotation of the Earth. Because of the Moon, the spin of the Earth is slowing down about one millisecond (0.0015 seconds) per year. About every eighteen months a leap second is added to keep our planetary time consistent with atomic clocks and astronomical observations. At the time of the dinosaurs, a day was about twenty-three hours long.

To make up for the slower rotation rate, the Moon is moving away from the Earth about 1.5 inches per year. As the time for rotation of the Earth grows longer, the Moon increases its orbital radius. It's that same figure-skater idea in reverse: slow rotation, arms out. We can think of the Earth and the Moon as being a system. The body of the skater is the Earth, and the arms are the Moon. As the skater (Earth) slows down, the arms (Moon) move away from the skater.

Millions of years from now, the Moon will look smaller in the sky and a day will be twenty-five or twenty-six hours. We'll all get more done!

65. What is the Bermuda Triangle, and why have people disappeared in it, never to be seen again?

Most rational explanations for incidents in the Bermuda Triangle involve pilot errors, a swift and turbulent Gulf Stream, and environmental factors—namely, weather.

The Bermuda Triangle is an area off the southeastern coast of the United States. The corners of the triangle are Bermuda; Miami, Florida; and San Juan, Puerto Rico. It's really an "imaginary" area, not recognized by the United States Coast Guard. But the area has become notorious for the seemingly high frequency of unexplained losses of ships, small boats, and planes. For example, the loss of five TBM Avenger airplanes (Flight 19) and the rescue aircraft sent after them, on December 5, 1945, is legendary.

However, the Bermuda Triangle is an area of heavy sea and air traffic covering a huge chunk of ocean. One would expect to have losses along such a heavily traveled route.

Another possible reason behind the Bermuda Triangle's link to disappearances is that compasses in the triangle don't point anywhere near to true north. Instead, they point to magnetic north, which is about eleven hundred miles away from true north. That magnetic variation is off by only a few degrees here in Tomah, Wisconsin, but it is off by nearly twenty degrees in the Caribbean. If navigators don't take that anomaly into account, they can be in deep trouble.

The Bermuda Triangle also happens to be a naturally danger-filled area. It falls within the course of the Gulf Stream, which is swift and turbulent. Storms can arise suddenly and unpredictably, which not only leads to accidents but also rapidly erases any evidence of plane crashes or shipwrecks. Most of the hurricanes that have made the news lately travel through the Bermuda Triangle. And waterspouts, which are tornadoes over water, are frequent. The terrain itself is also challenging to navigate. The ocean floor has some of the deepest trenches on the planet, but also extensive shoals and reefs. Many a ship has lost its bottom on those treacherous formations.

Lloyd's of London, which provides the insurance for most of the world's shipping, claims that losses are no higher in the Bermuda Triangle than in any comparable area of the world's oceans. Insurance rates are the same for Bermuda Triangle traffic as for ships that traverse any other area of the world.

Conspiracy theories abound in our media, running the gamut from Bigfoot and the Loch Ness monster to UFOs and aliens from outer space. These theories all sell books, magazines, and TV programs, but what they lack is credible evidence.

66. Why do rivers meander?

It would seem logical and natural for a river to run straight, but rivers that flow over gently sloping ground begin to curve back and forth. The rambling routes of what we call meandering rivers come about through erosion and sediment deposit.

Due to some asymmetry or obstruction in the riverbed, such as rocks, weed growth, or fallen trees, the speed of the flowing water between the two banks differs. The faster side of the river carries more sediment along, so less is deposited. And because the water is flowing faster, more erosion takes place. The slower side of the river deposits more sediment from erosion. Slower-moving water allows more time for soil particles to settle.

You can probably tell what is happening here. The faster-moving water is eating into the bank, making a small curve. Once the curve is established, the water on the outside of the curve must travel faster than that on the inside because of the greater distance the water must travel. This erodes the outside of the curve more, the water moves still faster, and the process perpetuates itself. As the river erodes soil from the outer curve, it deposits the sediment on the inner curve. This causes the meanders to grow larger and larger over time; the bend gets more and more pronounced. Consequently, the slower side of the river will continue to get slower and the faster side to get faster. Thus, more sediment gets deposited on the slow side and more erosion occurs on the fast side.

This process continues until the curve is so sharp that the river cuts through the bend and reestablishes a straight path. This can cut off a

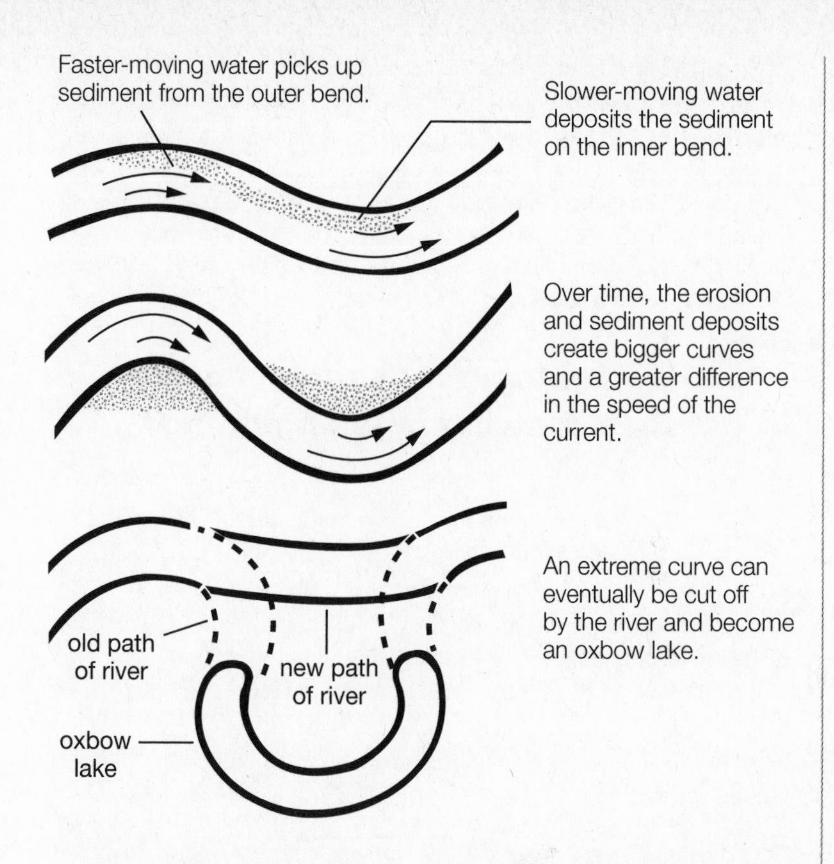

meander from the rest of the river, most often at a time of flooding. The cutoff part of the river forms an oxbow, named for the U-shaped part of the yoke for oxen, because the meander could be viewed as a pronounced U-shaped bend in the river. Over years, many oxbows fill in with sediments and plant growth.

From a high vantage point, one can see the oxbows from decades past. Some newly formed oxbows harbor water for years and form lakes if large enough. Carter Lake, Iowa, was created this way in 1877 after severe flooding shifted the Missouri River over a mile to the southeast.

The town of Horseshoe Lake, Arkansas, is built on the eastern tip of a U-shaped body of water with the same name, formed by changes in the course of the Mississippi River that created an oxbow. The lake is no longer connected to the Mississippi River.

The low-lying area around a river is termed a floodplain. Sediment is deposited in such an area after heavy rains and spring flooding, yielding some very rich land for crops. We had a few acres of that kind of farmland down on the farm I grew up on in Wisconsin. That soil was so rich that if you dropped a seed corn in a small furrow and covered it up with dirt, you had to get your head out of the way, lest the fast-growing stalk hit you in the face. Now, that is rich farmland!

67. Do lightning rods protect houses, or are they drawing in the lightning?

Ben Franklin gets credit for the invention of the lightning rod. A lightning rod is a rather simple device, a metal rod with a ball or a point at its end attached to a building's roof, chimney, or steeple. The one-inch-diameter rod is attached to a copper or aluminum wire connected to a conductive metal grid buried deep in the ground.

The purpose of the lightning rod is not to attract the lightning. It is actually supposed to discharge the clouds above it. Lightning is an electrical discharge between two points, namely cloud and ground or from cloud to cloud. You and I build up an electrical charge when we walk across a carpeted floor wearing socks and touch a metal doorknob. We also experience mini-lightning when we bring cotton and wool clothes out of the clothes dryer. The same thing happens when we pull a wool sweater off a cotton shirt we're wearing (see page 282).

Sometimes the electrical potential between the clouds and the ground is so strong that, despite the presence of a lightning rod, lightning strikes anyway. So the lightning rod provides a low-resistance path to the ground that can carry the tremendous current safely to the Earth.

Lightning is very finicky and can jump around. Lightning can strike and then seek a path of least resistance. Usually, but not always, it is the highest object closest to the clouds, such as a steeple or tower. In open areas, that could be nearby trees. Growing up on a farm, we would hear stories of lightning striking a grove of trees and killing cattle seeking refuge from a storm.

Lightning is nothing to fool around with. More than five hundred people get struck by lightning every year in the United States, and about a tenth of them die. The following are all bad ideas during a lightning storm: being out in the open, sitting in a boat on water, operating farm machinery, golfing, talking on the phone (although cordless and cell phones are okay), and taking a bath or shower.

Being inside a house away from windows is good. Sitting in your car (unless it's a convertible) or truck is ideal—just don't touch any metal parts that connect with the outside of the car. You're safe in the car because the lightning travels around the outside of the metal. It's called a Faraday cage. Contrary to popular belief, it is not the tires that save you.

In 1836, the English scientist Michael Faraday demonstrated that an electrical charge stays on the outside of a conductor. A metal box will keep electromagnetic radiation from penetrating the interior of that metal box. A box made of metal mesh or screen will accomplish the same thing, if the holes in the mesh are small compared to the length of the electromagnetic waves trying to get in.

There could be several reasons that your house, barn, or any structure, for that matter, is repeatedly struck by lightning. To protect it, the lightning rod wire must be attached securely to a wire cable. The wire cable should be buried as deeply as possible in the ground. But lightning is very unpredictable and capricious, so maybe you're just having bad luck!

Roy Sullivan, a US park ranger in Shenandoah National Park in Virginia, was struck by lightning seven times between 1942 and 1977. He holds the record. Two of the strikes even set his hair on fire. Strangely enough, it wasn't the lightning that killed him; he took his own life by a self-inflected gunshot wound in 1983, at age seventy-one, reportedly over a woman.

68. If heat rises, why is a mountaintop colder than the bottom?

At first thought, it might seem that that a mountaintop should be hotter than the lower valleys, because we know that hot air rises, and, additionally, mountaintops are closer to the Sun. The Sun beats down on the Earth, and air at the surface of the Earth is warmed.

The real reason is that, yes, warm air rises, but rising air expands. And as the rising air expands, it cools. Think of air molecules as tiny balls hitting against each other. A ball would pick up speed when another ball that is approaching hits it. But when a ball collides with another ball that is moving away or receding, its rebound speed is reduced. Or think of it this way. A Ping-Pong ball picks up speed when hit by an approaching paddle; it loses speed when hit by a receding paddle.

Higher up in the atmosphere, in the area of expanding air, molecules collide with more molecules that are receding than approaching. So the average speed of the molecules is decreasing. Hence, air is cooling because temperature is the average kinetic energy or velocity of molecules. How much does air cool as it rises? This is called the adiabatic cooling rate. It depends on the moisture in the air. On average, air cools about 3.5°F or 2°C for every one thousand feet in altitude.

The temperature on Mount Everest, altitude slightly less than thirty thousand feet, can be as low as –100°F. But on a good day in May, a climber can expect around –15°F.

Last summer, my wife, Ann, and I toured New Mexico and Arizona. Our rental car had a thermometer readout that gave the temperature of the air outside the car. When we were at Page, Arizona, altitude four thousand feet, the temperature was 92°F. A couple of hours later, we were at the North Rim of the Grand Canyon, altitude about nine thousand feet, and the temperature was 74°.

Arizona highways have signs that tell the altitude at every one-thousand-foot level. It was satisfying to keep track of the temperature versus the altitude along Highway 89 and then Highway 67 and watch the temperature drop about 3.5 degrees for every thousand-foot rise in elevation.

When we think of mountains in the United States, we think Rocky Mountains. The pioneers on the Oregon Trail in the mid-1800s would first glimpse the snow-covered peaks of the Rockies after they passed Independence Rock and started trudging up along the Sweetwater River.

What fear and trepidation they must have felt. But they were lucky. Lying before them was South Pass, discovered by fur trappers in 1812, a broad, open saddle of land with prairie and sagebrush, sort of a natural opening across the Continental Divide. South Pass is located in southwestern Wyoming at an elevation of seventy-four hundred feet. The Wind River Range is to the north, and the Antelope Hills are to the south.

Over a quarter of a million pioneers crossed the Rocky Mountains through South Pass. Today, Wyoming Highway 28 follows the Oregon Trail through South Pass. Wagon ruts are clearly visible at several places.

69. Why are the oceans salty?

The father of modern chemistry, Antoine Lavoisier, gave us the first definitive answer some two hundred years ago. He stated that oceans are the "rinsings of the Earth." He meant that salts are washed from the land into the ocean.

The rocks on land contain calcium carbonate (limestone), magnesium sulfate (Epsom salts), and sodium chloride (table salt). The process of weathering breaks down the minerals in these rocks and salts, and they dissolve in the water as rivers and streams wash the salts from the land into the ocean. We can see evidence of this interaction of water and stone at most any cemetery. Many early headstones were made of marble. After a hundred years of wind and rain, the inscriptions are hard to read.

The salt in the ocean also comes from volcanic activity. While there is hardly any sulfur and chlorine in rocks, volcanoes spew these elements into the atmosphere, and they end up falling in the world's oceans. Sulfur and chlorine add to the saltiness of the oceans.

Since weathering and volcanic eruptions continually happen, it might seem that the oceans should become more and more salty. But salt is constantly being removed by clams and other shellfish that use calcium carbonate to build their shells. So the salinity of the oceans has remained fairly consistent for a long time.

The Dead Sea, on the border of Israel and Jordan, is surrounded by the lowest land on Earth, and it's also one of the world's saltiest seas. It has a salinity of 31.5 percent, almost nine times that of the ocean. The Dead Sea has no outlet. The minerals that flow into the Dead Sea stay there for centuries. The majority of freshwater bodies have rivers flowing out of them that dispose of dissolved minerals. Not so for the Dead Sea.

Mono Lake, near Yosemite National Park in California, has a salinity of about 10 percent. Diverting of water to Los Angeles caused the high salt content, leading to loss of water due to evaporation that soon exceeded freshwater inflow from streams. The lake continued to get smaller and saltier until 1994, when the drawdown was finally halted.

The Great Salt Lake in Utah is the remains of a pluvial, or rainwater-filled, lake that covered much of Utah in prehistoric times (see page 78). Today, three rivers deposit their sediment in the lake; they leave behind more than a million tons of minerals each year. The lake has no outlet, so water disappears by evaporation only. When water evaporates, the minerals are left behind. No need to worry about drowning in the Great Salt Lake; people float in it, because the water is denser than the human body.

70. Why is the Earth round?

All the planets are round because of gravity. Gravity is the force exerted inward toward the center of the planet so that all parts of the surface are pulled evenly toward the center. The result is a sphere, or ball. When the Earth and other planets were forming, gravity gathered all the gas and dust into bigger and bigger clumps. Collisions made the material hot and molten, and gravity pulled it all inward as much as possible to make a sphere. Later, all the molten material cooled and hardened in these spherical forms.

Planets are not perfect spheres. Any spinning mass wants to throw that mass to the outside, as far away from the center of rotation as possible. Witness the shape of a lasso, or lariat. A planet's rotation causes it to bulge out more at the equator. The Earth is close to being a sphere, but not quite. The bulge from its spinning makes it twenty-six miles farther from one point to the opposite point on the equator than it is from pole to pole.

Jupiter is flattened, or oblate, by about 7 percent. It is a greater distance around the equator of Jupiter than it is around the poles. Jupiter is a huge planet, with about 320 times the mass of Earth. It spins once on its axis in about ten hours. You can look at Jupiter through a small telescope and see that it is not spherical. And you can use a NASA photo of Jupiter to measure the difference between a polar diameter and equator diameter.

Mountains and valleys, on planets that have them, are smaller deviations from a perfect sphere. The highest mountains on Earth are about six miles high, and gravity prevents them from getting much higher. Mountains that were fifty or one hundred miles tall would be crushed by their own weight.

71. If there is no air in space, how is there air on our planet?

Early on in our planet's history, there were millions of volcanoes, with steam belching out of their mouths. Steam is a form of water, which has two hydrogen atoms and one oxygen atom. These volcanoes also emitted carbon dioxide and ammonia. Carbon dioxide has one carbon atom and two oxygen atoms. Ammonia has one nitrogen atom and three hydrogen atoms. The elements of these gases provided the building blocks of the air we breathe today.

Carbon dioxide levels in the atmosphere dropped because most of the carbon dioxide dissolved into the oceans and simple bacteria took in carbon dioxide, along with sunlight. These bacteria produced oxygen as a waste product, causing oxygen to build up in the atmosphere. Sunlight broke apart the ammonia molecules, separating them into nitrogen and hydrogen. Hydrogen is the least dense of all the elements, so it drifted off into space. A hydrogen atom has a large amount of kinetic energy, the energy of motion. It escapes the Earth's gravitational pull and heads for outer space. Hydrogen atoms can exceed the Earth's escape velocity of 25,000 mph. The nitrogen from the ammonia remained and eventually became the predominant gas in the planet's atmosphere.

These days, plants and animals thrive in a delicate balance. Animals take in oxygen and give off carbon dioxide. Plants take in carbon dioxide and emit oxygen. Life depends on this delicate equilibrium of 78 percent nitrogen, 21 percent oxygen, and 1 percent is carbon dioxide and the inert gases of argon, krypton, neon, xenon, and helium. These gases were present in minute amounts after the formation of the solar system, some 5 billion years ago.

Our Earth has oxygen, and that's what we humans need to sustain life.

Earth's atmosphere protects us from dangerous radiation, especially potentially deadly ultraviolet radiation. The atmosphere also keeps us warm by holding heat close to the surface of the Earth and not letting it escape and provides us with rain that nourishes and irrigates the planet.

Some other planets in our solar system have atmospheres, too, but they can't support life. Mercury, like the Moon, has no atmosphere. Venus is cloaked in dense carbon dioxide gas. We humans would have a hard time finding oxygen to breathe. Mars has an atmosphere that has one-hundredth the density of our own, very close to a vacuum. The large planets Jupiter, Saturn, Uranus, and Neptune have atmospheres of hydrogen, helium, methane, and ammonia.

It does appear that Earth is the only planet in our solar system that is comfortable to live on.

72. What prevents skyscrapers from sinking into the ground?

Very tall skyscrapers must be built on bedrock. The chief obstacle to building upward is the downward pull of gravity. Every time one adds more floors on top, the total force on the bottom layers increases. You could build upward almost indefinitely, but the bottom floors would eventually have to be so massive and so thick that there wouldn't be any living space left inside.

Each vertical column in a skyscraper sits on a spread footing: the column sits on a cast-iron plate, which sits on top of a grillage. The grillage is a spread-out stack of horizontal steel beams that sits on a thick concrete pad, which sits on bedrock. Bedrock is the hard, consolidated, intact material that can support a massive weight.

New York City's Manhattan sits on the hardened remains of molten lava that flowed down the Hudson Valley centuries ago. The skyline of Manhattan traces out the subterranean mountain range of that lava flow, including a fold in the lava toward the southern end of the island.

Lower Manhattan has skyscrapers like the new One World Trade Center and the Woolworth Building. There are no tall buildings stretching more than a mile north from there. In midtown Manhattan, we once again see tall buildings: the Empire State Building, Rockefeller Center, the UN Building, and the Chrysler Building. Tall buildings are built on the two underground mountains where the lava flow is close to the surface.

The tallest building in the world at the present time is the Burj Khalifa, at 2,722 feet. Located in downtown Dubai in the United Arab Emirates, the South Korean–constructed building opened on January 4, 2010.

73. Why does it rain?

Rain is the result of the continuous water cycle. Water evaporates from lakes, oceans, rivers, and wet ground. To "evaporate" means to go from a liquid state to a vapor state. As the warm moisture-laden air rises, it expands because the air it is rising into is less dense (see page 76). That expansion lowers its temperature, because air molecules don't collide with each other as frequently as they do in warmer air. The cooler air cannot hold as much moisture in vapor form as warm air can hold. The relative humidity shoots up and eventually reaches 100 percent. Relative humidity is the percentage of the moisture in the air compared to the maximum amount of moisture the air can hold.

The temperature at which the air has cooled enough to be completely saturated is termed the dew-point temperature. The cool air can no longer hold water vapor and gives it up; it condenses (goes from vapor to liquid state) on microscopic particles such as pollen, dust, smoke, and oxygen molecules in the atmosphere, forming a cloud (see page 116). These tiny water droplets in the clouds fuse together, or coalesce, to form larger water droplets. Air turbulence and winds move these droplets around, making them bigger as they collide. Soon they become heavy enough to overcome rising air currents, and they fall as raindrops.

In special conditions, such as a cumulonimbus thunderstorm, the raindrops are driven upward by rising air currents to where the temperature is below the freezing point. Tiny ice balls fall and gather more raindrops. The ice balls may ride up and down the air currents, and each time they do, they gather more water on their surfaces. Each time they rise, they freeze. Finally, they're heavy enough to fall all the way to the ground. We call this hail. Next time it hails, go outside and gather a few big hailstones and take them in the house. Use a parry knife and cut one in half and examine the cross-section. On a marble-size hailstone, you can see and count the layers of ice, much like determining the age of a tree by counting the tree rings. The rings of ice will tell you how many times the hailstone rode up and down in the storm clouds.

The driest habitable place on Earth is Arica, in the Atacama Desert of Chile, where the annual rainfall is .03 inches. The cactuses take moisture from the fog.

The wettest place on Earth is claimed by both Llora, Colombia, in South America and Cherrapunji, in northeast India. Llora receives 40 feet (yes, feet) of rain per year.

The heaviest rainfall recorded was in Holt, Missouri, where on June 22, 1947, twelve inches of rain fell in forty-five minutes. On July 3, 1976, ten inches of rain fell in four hours over Big Thompson Canyon, Colorado, which is on the way to Estes Park and Rocky Mountain National Park. Flash flooding killed 144 people, mostly campers.

Despite the damage it can sometimes cause, rain is a blessing. It is the primary source of freshwater the world over, and it warms and irrigates our good Earth. Rain makes life and all human activity possible.

74. What is the hardest material on Earth?

For hardness, diamond is the standard by which all other materials are judged (see page 82). Mention diamonds and most of us think engagement rings, anniversary rings, and "a girl's best friend." But those in industry value diamonds for their use as cutting tools, abrasives, and wear-resistant protective coatings.

Diamonds are a form of carbon, which is one of the most common elements in the world and one of the four essentials for the existence of life, the others being water, food, and oxygen. We humans are more than 18 percent carbon, and the air we breathe contains traces of carbon. Diamond formation takes place about one hundred miles below the surface of the Earth, where the tremendous heat and pressure change carbon into diamonds. The majority of the diamonds we see today were formed billions of years ago and brought to the surface of the Earth by magma eruptions. Most of the huge and more famous diamonds were found in South Africa.

The Mohs Scale is used to determine the hardness of solids, especially minerals. It is named after the German mineralogist Friedrich Mohs. The scale reads as follows, from softest to hardest:

1. **Talc**—easily scratched by the fingernail
2. **Gypsum**—just scratched by the fingernail
3. **Calcite**—scratches and is scratched by a copper coin
4. **Fluorite**—not scratched by a copper coin and does not scratch glass
5. **Apatite**—just scratches glass and is easily scratched by a knife
6. **Orthoclase (feldspar)**—easily scratches glass and is just scratched by a file
7. **Quartz (amethyst, citrine, tigereye, aventurine)**—not scratched by a file
8. **Topaz**—scratched only by corundum and diamond
9. **Corundum (sapphire and ruby)**—scratched only by diamond
10. **Diamond**—scratched only by another diamond

As this scale indicates, no other material can scratch diamond, making it the hardest *natural* material on Earth. However, recently a patent was taken out for a *synthetic* compound of carbon and nitrogen that is said to rival diamond in hardness. It will serve as an inexpensive substitute for diamonds used in industrial applications. This new super-hard material could be used to cut steel, which diamond can't do because it burns when it gets hot. Also, this new synthetic "diamond" might be used to coat metals such as gears and bearings and make them last a lot longer. There are two kinds of these synthetic diamonds, one going by the name of high-pressure high-temperature (HPHT) diamonds and the other by chemical vapor deposition (CVD) diamonds.

75. How far is it to the bottom of the Pacific Ocean?

The deepest place in the oceans is the Marianas Trench in the Pacific Ocean. In 1960, the US Navy sent the *Trieste,* a mini-submersible, or bathyscaphe, named for the city in Italy where it was built, down to the bottom. A bathyscaphe is a hollow metal ball-shaped diving vessel that can hold a small crew and is used to explore ocean depths. Unlike earlier bathyscaphes, the *Trieste* was not tethered to the mother ship above it.

The two-man *Trieste* crew consisted of Jacques Piccard, son of the vessel's designer, Auguste Piccard, and US Navy Lieutenant Don Walsh. They went down roughly thirty-six thousand feet and rested their vessel on the bottom of the ocean floor. It took them four hours to descend to the bottom of the deepest place on Earth. More than seven miles of ocean were above them. The *Trieste* used a tank filled with gasoline, which is lighter than water, and lead pellets, which are heavier than water, as ballast to control buoyancy. The pressure of the water at the lowest point was sixteen thousand pounds per square inch. In comparison, the air pressure in a car tire is a tad over thirty pounds per square inch.

The *Deepsea Challenger* is a modern, small, high-tech submarine that carries a one-person crew. On March 26, 2012, Canadian film director James Cameron piloted his craft to one of the deepest parts of the Marianas Trench, a recorded depth of 35,756 feet. This depression, the Challenger Deep, was named in honor of the British Royal Navy survey ship HMS *Challenger*, whose crew made the soundings in its 1872-to-1876 voyage. Cameron spent three hours exploring the ocean bottom in his solo dive.

Cameron stated that he saw no fish at the bottom of the ocean, only small amphipods, which are shrimplike bottom feeders. Rolex was a major sponsor of the vessel and voyage. They wrapped a Rolex watch on the vessel's robotic arm. It functioned normally throughout the dive. Very good advertising material!

76. What keeps the Earth in its orbit?

Two phenomena are operating at the same time to keep Earth in its orbit. First, the gravitational pull of the Sun is pulling the Earth toward it. Second, the forward motion (inertia) of the Earth is guiding the planet in a straight line. It is the competing action of these two "forces" (inertia is technically not a force, but it's responsible for the planet's trajectory because the Earth "wants" to go in a straight line, at a tangent to its orbit) that makes the Earth go in a smooth, and nearly circular, orbit around the Sun.

If only one of these "forces" were acting, it would be a disaster for the Earth. If our world were somehow made stationary in its path, the Sun's gravity would pull it in and burn it to a cinder. On the other hand, if the Sun's gravity somehow magically disappeared, Earth would fly off tangent to its orbit and be lost forever.

It is much like swinging a ball on the end of an elastic string around your head. If the string broke, the ball would go flying off in a straight line. The string is much like the gravitational pull between the Earth and the Sun. If the ball suddenly stopped moving forward, the elastic string would pull the ball toward your hand.

77. What are the special conditions on Earth that allow it to support life?

Anyone who has ventured out in the country and gazed up into the night sky has asked the question, "Are we alone in the universe?" It is a difficult question, for it is one for which we have no data or proof. Earth is the only place in the universe on which we are certain that life exists.

Now, we have to define what we mean by life. There is general agreement that life forms have the following four characteristics: 1. They can react to their environment and can heal themselves when damaged. 2. They can reproduce and pass on some of their characteristics to their offspring. 3. They can grow by taking in sustenance from the environment and changing it into energy. 4. They have the capacity for genetic modification so as to adapt to a changing environment over the course of generations.

The educated guess of most astronomers and cosmologists is that the universe is teeming with life. They argue from the laws of probability, and their hypothesis goes something like this: Earth is one little planet orbiting one little ordinary star. The universe is filled with billions of galaxies, with billions of solar systems and billions of planets going around them. Why should Earth be the only planet among those billions on billions that has life?

There is another case to be made for extraterrestrial life. Life on Earth depends on just a few basic molecules. The elements that make up those molecules are common to all stars. The laws of science apply to the entire universe. Given sufficient time, life must have originated elsewhere in the cosmos.

An opposing view maintains that intelligent life on Earth is a result of a whole series of extremely fortunate accidents—geological, astronomical, chemical, and biological—and therefore life anywhere else in the universe is unlikely.

There is yet a third view, the creationist point of view, that claims that life was created, resulting from the actions of a supreme being. This is a religious point of view that says God created humans out of nothing and that any changes in life forms are divinely directed.

Will we ever find out? If in some distant era we hear the faint but unmistakable radio chatter of an advanced civilization, it will be a profound moment in the history of humanity, for then we will know that we are not alone. If we never hear anything, that will also be profound. It will mean we are unique and there is nothing like us in the universe!

78. How do we know the Earth's age?

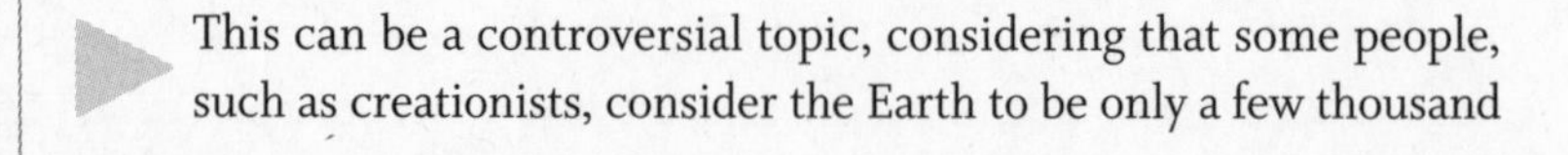

This can be a controversial topic, considering that some people, such as creationists, consider the Earth to be only a few thousand years old. But the age of the Earth is firmly rooted in solid scientific logic and empirical evidence. The Earth and the solar system were formed between 4.53 and 4.58 billion years ago. Our Milky Way Galaxy is roughly 13 billion years old.

What's the evidence? The ages of Moon and Earth rocks are found by measuring the decay of long-lived radioactive isotopes. Radioactive uranium-235 and uranium-238 undergo a change or transmutation (decay) into other elements, eventually ending up as stable lead. The more plentiful U-238 yields Pb-206. Pb-207 is the end result of the rarer U-235. All the lead that we have on Earth came from uranium.

Let's say you have a rock that has a high ratio of uranium to lead—in other words, a lot of uranium and not much lead. This means you have a very young rock, because it has not existed long enough for much of the uranium to change to lead.

Now let's pretend you're holding a rock that has a low uranium-to-lead ratio: a lot of lead, but not much uranium. This means you've got yourself a very old rock, one that has lasted a sufficient time for most of the uranium to change to lead. Determining the age of the Earth is harder than finding the age of the Moon. Earth rocks have been recycled and destroyed by plate tectonics, uplifting, heating, and cooling. It's hard to find rocks on Earth that are in their original, unaltered state, but rocks exceeding 3.5 billion years of age have been found on every continent.

The Moon is another story. Not many changes have occurred on the Moon in billions of years. While the oldest rock found on Earth has been dated at 4.3 billion years, the oldest Moon rock brought back by the astronauts was shown to be between 4.4 and 4.5 billion years old.

The best estimate of the age of the Earth comes not from dating individual rocks but instead from considering the Earth, the Moon, and meteorites as part of the same system, in which the ratio of uranium to lead can give a quite precise age. Thousands of meteorites, of over seven different types, have been radiometric dated. They show an average age of 4.56 billion years. Taken together with the ages of the oldest Earth and Moon rocks ever found, we arrive at our estimate of 4.53 to 4.58 billion years for the age of the Earth.

79. How is dirt made?

Dirt is the thin layer of soil that covers our planet. In most places, it is just a few feet thick, because nearly all of the Earth is a big, hard, solid rock, with an inner liquid core.

Dirt is mostly made of bits and pieces of this rock, which is broken down into smaller and smaller pieces because of weathering and microorganisms breaking down plant matter. Moisture, temperature, wind, rain, freezing, and thawing are all part of the weathering process. Over hundreds of years, rocks break down into tiny grains, and these small grains, mixed with plant and animal matter—decayed roots, leaves, dead bugs and worms, and other organic matter thrown in, along with water and air—is what we call dirt or soil.

The type of rock determines the alkalinity and texture of the soil. Limestone produces soils that are fertile, neutral (not base or acid), and finely textured. We have a lot of limestone-based soils where I'm from in Wisconsin. Soft shale rock yields a heavy clay soil. Sandstone becomes a coarse, sandy soil. Granite gives a sandy loam and acidic soil.

Dirt that is dark and black has a lot of old plants in it. The dark soils of southern Minnesota are some of the richest on Earth. Dirt that is light-colored contains a lot of silicate, or sand. Sandy soils drain quickly and tend to need a lot of water to grow productive crops, which is why you often see irrigation systems on land with sandy soil.

Clay soils are composed of extremely fine minerals and flat particles that pack together tightly. Clay soils tend to be reddish, harden when dry, and drain poorly. They tend to feel sticky when wet. The southern states of Georgia and Alabama are examples of areas with clay-based soils.

On a lighter note, dirt is what you track in on your mother's floor. Soil is what plants grow in. Dirt is what you get on your uniform sliding into second base. Soil is vital to the crops that feed people. Every state has selected a state soil, twenty of which have been established by their states' legislatures. In my home state of Wisconsin, the 1983 legislature named Antigo silt loam as the official state soil. It is a well-drained soil suited for forests, dairy, and potatoes.

80. Why does the Earth have gravity?

We are all familiar with gravity. We know how it behaves. Hold a ball in your hand and let go; sure enough, it drops straight downward. And we all have an innate fear of gravity. Get on top of a tall ladder or tall building, look down, and we instinctively know what gravity will do if we lose our balance.

Gravity makes falling objects go faster and faster. In other words, objects accelerate when they fall. In the English system of measurement, a falling object goes thirty-two feet per second faster after each second. That works out to about 22 mph faster each second. In the metric system, it's about ten meters per second faster than the previous second. These numbers are called "the acceleration due to gravity."

The force, or pull, of gravity depends on the mass of the Earth and of the object being pulled. That force of gravity on Earth is known as the weight of the object. The more mass you and I have, the more we weigh.

Gravity gets weaker the farther you get from the surface of the Earth. Like many forces in nature, it obeys the inverse square law. Twice as far from the Earth, the pull of gravity is the inverse of two squared, or one-fourth. Three times greater distance means the inverse of three squared, so gravity's pull is one-ninth.

Isaac Newton, beginning with experiments in 1667, was the first to get a mathematical handle on gravity. He figured the same gravity that caused an apple to fall from a tree was the force that made the moon "fall" around Earth.

Albert Einstein, in the early twentieth century, redefined gravity as a sort of geometry of space and time. The presence of mass causes a curvature of space-time. So objects follow a curved path, and their movement along this path we call "acceleration."

Still, gravity is a great mystery. Why should any two objects in the universe attract each other? One goal of science today is to combine several forces, namely gravity, electromagnetism, and two nuclear forces, into a single unified theory that answers such questions. For now, the simplest and best explanation of why the Earth has gravity is that the Earth has mass and gravity is a property of mass.

81. What creates the wind?

Wind is caused by a difference in pressure from one area to another area on the surface of the Earth. Air naturally moves from high to low pressure, and when it does so, it is called wind.

Generally, we can say that the cause of the wind is the uneven heating of the Earth's surface by the Sun. The Earth's surface is made of different land and water areas, and these varying surfaces absorb and reflect the Sun's rays unevenly. Warm air rising yields a lower pressure on the Earth, because the air is not pressing down on the Earth's surface, while descending cooler air produces a higher pressure.

But there are many other factors affecting wind direction. For example, the Earth is spinning, so air in the Northern Hemisphere is deflected to the right by what is known as the Coriolis force. This causes the air, or wind, to flow clockwise around a high-pressure system and counter-clockwise around a low-pressure system. The closer these low- and high-pressure systems are together, the stronger the "pressure gradient," and the stronger the winds. Vegetation also plays a role in how much sunlight is reflected or absorbed by the surface of the Earth. Furthermore, snow cover reflects a large amount of radiation back into space. As the air cools, it sinks and causes a pressure increase.

And wind can get even more complex. Some parts of the Earth, near the equator, receive direct sunlight all year long and have a consistently warmer climate. Other parts of the Earth, near the polar regions, receive indirect rays, so the climate is colder. As the warm air from the tropics rises, colder air moves in to take the place of the rising warmer air. This movement of air also causes the wind to blow. It's a dynamic, complex mechanism, which is why weather forecasting is not quite a precise science.

Today we see windmills, used to make electricity, in operation in all parts of the United States, but especially along our coasts. Coastal regions tend to have fairly strong winds blowing in from ocean to land during the day and out from land to ocean during the night. The cause of this phenomenon is that land heats up and cools down faster than water, again creating a pressure gradient.

3

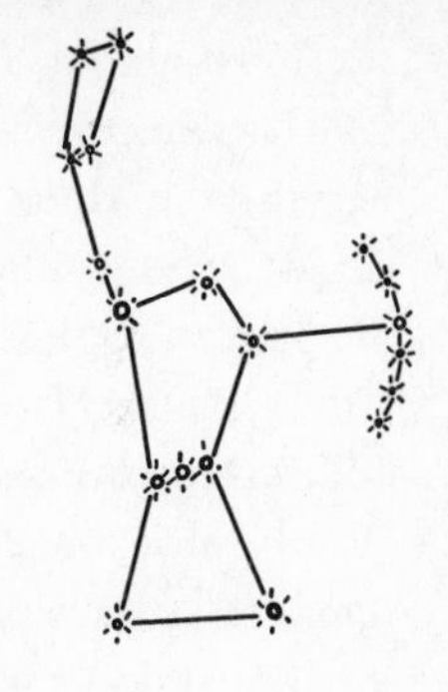

Science in the Sky

82. Why do we sometimes see the Moon in the daytime?

There are two factors behind this phenomenon. First, the Moon is very bright. In fact, it is so bright that you can see it against the blue backdrop of the daytime sky. It is not as bright as the Sun, of course, because the Moon doesn't give off any light by itself. The Moon only shines by reflecting light from the Sun. However, the Moon is one hundred thousand times brighter than the brightest nighttime star.

The Moon reflects so much light that our farming ancestors used its light to keep working to bring in the crops well into the nighttime hours. Hence, they named the full moon closest to the autumnal equinox (September 21 or 22) the harvest moon. Once every four years, the harvest moon is actually in October. The full moon after the harvest moon is referred to as the hunter's moon.

The second idea to consider is that the Moon goes around the Earth in a little less than thirty days, or what we call roughly a month. For part of that time, around the time of the full moon, the Moon is in the sky opposite the Sun. The Moon is rising as the Sun sets. But on each successive night the Moon comes up about an hour later. The Moon gets closer and closer to the direction of the Sun in the early-morning hours, finally appearing very close to the Sun in the sky around new moon time, rising and setting at about the same time as the Sun.

Then the Moon gets farther and farther from the Sun in the sky each night until it gets back to full moon. The portion of the Moon that is lit up depends on the angle the Moon makes between the Earth and the Sun, which causes the various phases of the Moon.

During these phases, people on Earth can see the sunlight that strikes the Moon and is reflected from the Moon to us on Earth. So then we can see the Moon during the day. Again, if you see the Moon rising after sunrise, it is going from new moon to full moon, so each morning you see more and more Moon.

The Moon is "out" as many hours during the daytime as it is "out" during the nighttime. It's just that we associate the Moon with the nighttime sky. Of course, it is brighter in the night sky and, hence, more noticeable.

We use the Latin word "*luna*" to refer to anything pertaining to the Moon. "Luna" is a Latin word etymologically related to "*lucere*," which means "to shine." Now, this "moonshine" is not the same as the illegal liquor by the same name.

Get a softball, a basketball, and a lamp. The lamp is the Sun. Face the lamp and hold the basketball (Earth) in front of it. As you move the softball (Moon) around the basketball, you will be able to see all the phases, and you will also see that the Moon is visible during the daytime hours. Mentally and physically put yourself in place of the basketball. As the softball Moon goes around your head, you will be able to see the Moon in the "daytime." Remember that daytime for you is when your eyes can see the Sun.

83. What will happen to the Earth if the ozone layer keeps ripping?

We do need that ozone layer. It acts as a very thin shield high in the sky that protects us from the Sun's harmful ultraviolet (UV) rays. Bad things can happen if too much UV light reaches the Earth's surface: skin cancer and possibly eye damage (especially cataracts) and the weakening of our immune system, to name a few. Harm to our immune system means we have a harder time fighting off diseases.

Most of the ozone is in a layer that starts six miles above the surface of the Earth and goes up to about thirty miles above the Earth's surface. The ozone layer is very good at blocking UV-B rays, which are the rays that cause sunburn and most skin cancers.

In the 1970s, scientists noticed that some of the ozone was going away, or being depleted. The main culprit was chlorofluorocarbons, or CFCs, which are used to keep things cold and to make foam and soaps. Some of these "bad guys" were released in the air, mainly from fire extinguishers, aerosol spray cans, manufacturing facilities, refrigerators, air-conditioning units, and dry-cleaning establishments. CFCs are very stable at low altitudes, so they stay in the atmosphere long enough to diffuse into the stratosphere. Up there, UV rays are strong enough to break them down and release chlorine. Each chlorine atom can attack and break apart thousands and thousands of ozone molecules. So just a little bit can do major damage.

Countries around the world knew that CFCs were really bad stuff; scientists noted as early as the 1970s that there was a seasonal depletion of ozone over the Antarctic continent. Little worldwide action was taken until 1987, when an international treaty, called the Montreal Protocol, sharply limited the production of CFCs. They were completely phased out in the developed world by 1996. A hole in the ozone layer over the Antarctic was discovered in 1985.

Three satellites and three ground stations have shown that the depletion of the ozone layer is decreasing. Nature is repairing the damage, and ozone levels started to rise ever so slightly in late 2009 in most parts of the world. It will be fifty years before all the damage is repaired. Humans have the capability to do damage to our planet Earth, but we also have the knowledge to protect our home, if we'll use it.

84. What are the white lines that you often see behind airplanes?

Those thin-line clouds trailing behind jet aircraft are condensation trails, often shortened to "contrails." Jet fuel, which is low-grade kerosene, is made up of carbon and hydrogen. When jet fuel burns with oxygen in the atmosphere, most of the exhaust consists of carbon dioxide (CO_2) and water vapor. Generally, the water vapor is invisible. But cold air in the upper atmosphere can't hold nearly as much water vapor as warm air can. So the water vapor condenses onto tiny particles, such as exhaust particles and dust in the air, and forms the clouds we see.

We create the same phenomenon when we go outside in the winter and exhale. We see our own breath, which contains a good deal of that invisible water vapor. The breath hits the cold air, and the water vapor condenses onto tiny particles in the air. We don't see our breath in the summer because the warmer air holds the water vapor.

Contrails were noticed back in the 1940s, especially when the US Eighth Air Force sent hundreds of B-17 bombers from England to hit targets in Nazi Germany. Americans back home saw the contrails on Movietone newsreels seen in movie theatres.

The length of time that contrails persist depends on the altitude, temperature, water vapor content, pressure, sunlight, and shear winds. If the winds aloft are calm, the contrail will keep its shape and be seen from horizon to horizon. Sometimes, strong winds will spread out the contrail so it looks like those high-altitude cirrus clouds, often referred to as "mare's tails."

There has been speculation in some quarters concerning whether contrails are causing weather modification, a concept that isn't entirely science fiction. Cloud seeding has been in operation in various locations around the globe since the 1950s. Shooting silver iodide crystals into clouds causes the clouds to give up their water content. The water vapor condenses onto the crystals. Cloud seeding has been used to relieve drought, increase snowfall, dissipate hurricanes, and suppress hail. The Chinese government promised clear skies for the August 8 opening of the 2008 Summer Olympics. They launched 1,100 rain dispersal rockets from twenty-one sites around Beijing. Sure enough, no rain fell on their parade.

People in rural areas tend to pay more attention to the skies. They generally report seeing more contrails these days than they did decades ago. That is no doubt true, as there has been a large increase in jet traffic, both civilian and military. In addition, much of that air traffic has been a higher altitudes, some as high as forty thousand feet, where winds aloft are less likely to disperse the contrails quickly.

There have been those stories, held by conspiracy theorists, that chemicals are being spread from planes for a certain purpose. They call them "chemtrails" rather than "contrails." Some of the chemtrail conspiracy stories that float around the Internet claim that barium, aluminum salts, thorium, and silicon carbide are being released. Other accounts have the skies being seeded with electrically conductive materials as part of a super-weapons program. Other reasons stated are population control and alleviating global warming. Studies done in the United States, Canada, and Great Britain found no scientific evidence to support the allegation that high-altitude spraying is being conducted.

These operations may very well be going on, but thus far nobody has brought forth any proof or evidence. They remain, for now, conspiracy theories.

85. Why does the Moon have craters?

Our Moon's surface bears the marks of many craters formed by collisions from meteorites, comets, and asteroids. The average speed of a body striking the Moon is about twelve miles per second. When one of these visitors strikes the solid surface of the Moon, the resulting shock wave fractures the rock and digs a cavity, or bowl-shaped hole, about ten to twenty times the diameter of the impacting rock. The collision shatters the asteroid into smaller pieces and generates heat that may either melt or vaporize them.

The Earth has many craters from collisions with asteroids and meteorites. But the Earth has a thick atmosphere that acts as a shield. As soon as an asteroid comes into contact with our atmosphere, the air in front of the asteroid packs together, and the resulting friction increases the temperature to thousands of degrees. The meteorite catches fire. We see it as a shooting star, or falling star, as the burning particles trail behind the meteorite. Most of these meteorites disintegrate before they have a chance to reach the surface of the Earth. Our atmosphere acts as a safety shield and a cushion to protect the surface we live on.

The Moon has no such protecting atmosphere. Scientists estimate that over a ton of meteorites hit the Moon every day, beating it up rather badly. The surface of the Moon reveals the evidence of millions of years of bombardment. Copernicus is a large crater, sixty miles across. On Earth, plate tectonics, wind, rain, glaciers, and surface changes have eroded most of the impact craters away. Similar processes do not exist on the Moon, so its craters have remained almost unchanged over billions of years. About the only thing that erases a crater on the Moon is a new impact that leaves a bigger crater.

The Earth actually has more impact craters than the Moon. The Earth's diameter is four times that of the Moon, so it presents a bigger target for wayward asteroids and comets.

All the solid bodies of the solar system exhibit impact craters. Mercury looks much like the Moon. Venus's atmosphere consists of thick carbon dioxide clouds, but remote robotic landers show that its surface is heavily cratered. Venus and Mars have had volcanic activity that filled in many of their craters.

There are fifty-seven known impact craters in North America alone, with more than 170 spread over the entire Earth. The Chicxulub crater in the Yucatán Peninsula of Mexico is not easy to see, but satellite images, local changes in the gravity field, and ringlike structures in the land around the impact site give clues to its size. Many scientists believe the resulting fires, tsunamis, and clouds of dust and water vapor contributed to the extinction of dinosaurs 65 million years ago. The oldest and largest impact crater recognized on Earth is the Vredefort crater in South Africa. It is two billion years old and one hundred miles across. A visit to the Barringer Crater near Winslow, Arizona, must go on your bucket list. A 150-foot-wide iron-rich meteoroid struck there fifty thousand years ago. It left a crater about three-fourths of a mile wide and 750 feet deep. The site has a beautifully constructed visitor center, extensive displays, and guided tours. You can walk down into the crater or all the way around it if you are so inclined. In 1908, a large meteoroid or comet hit in Siberia, Russia, near the Tunguska River. This "Tunguska event" was a powerful explosion, equivalent to 185 of the A-bomb dropped on Hiroshima, Japan, in 1945. The blast knocked over 80 million trees and killed a lot of reindeer. Recently, on February 16, 2013, a fifty-five-foot-wide rock lit up the skies over the Ural region of Russia. It was traveling at 40,000 mph. It broke up at five miles above the earth's surface. The shock wave injured roughly 1,200 people.

86. How were the Moon's phases named?

The Moon has eight phases. When we see a big disk in the sky, that's a full moon. It looks like a shiny quarter or perhaps a twenty-dollar gold piece. When we can't see the Moon, that is the new moon phase. When the Moon appears to be getting bigger, it is waxing. When the Moon is getting smaller, it is said to be waning. When the Moon appears like the edge of a fingernail or a backward letter "C," just coming out of its new moon phase, it is in the waxing crescent phase. Over several days, as more of the Moon is lit up, it reaches the first quarter, which is when it looks like half of the Moon is lit and resembles the letter "D."

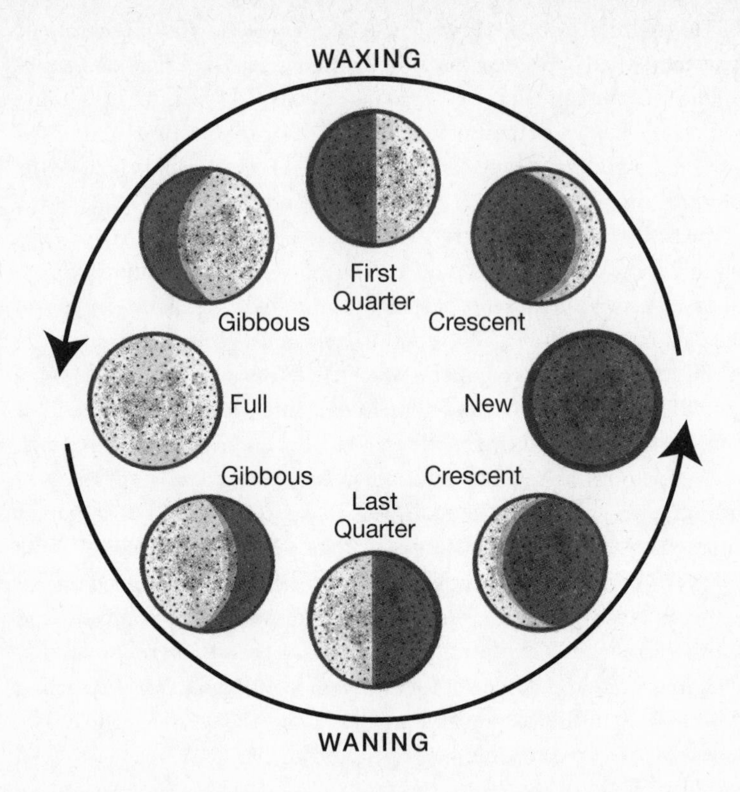

The term "first quarter" means that it has moved a quarter, or one-fourth, of the way from one new moon to the next new moon. After first quarter, the Moon seems to grow more and get bigger; we call this the waxing gibbous phase. The name "gibbous" comes from the Latin word for "hump"; in this phase the Moon has sort of a humpback shape to it. The halfway point of the Moon's cycle is the full moon, after which the Moon shrinks to waning gibbous, then last, or third, quarter, followed by waning crescent, then back to new moon.

Our Moon is not the largest moon in the solar system. Ganymede, a moon of Jupiter, is thirty-three hundred miles across, much larger than our Moon, which is twenty-two hundred miles in diameter. But our Moon is the largest moon in proportion to the planet it orbits. Because of its large comparative size, our Moon greatly affects our planet. The Moon is responsible for the tides (see page 129), and it also causes the Earth to weave back and forth in its orbit. The Moon's tug on the Earth makes the Earth have one big wobble every twenty-six thousand years. This causes the Earth's axis to point to different stars in the sky over the millennia.

"*Luna*" is the Latin word for month, which defines the time it requires the Moon to make one revolution, or orbit, around the Earth. The Moon is 240,000 miles from the Earth. It looks the same size in the sky as the Sun, but the Sun is four hundred times farther away. It is this same apparent size that allows us to see lunar and solar eclipses.

The Moon is about 4.5 billion years old, same as the Earth; they were formed at the same time. The Moon has no air or atmosphere, so its surface temperatures vary widely. Where it faces the Sun, the surface is 250°F. The part of the Moon's surface that faces away from the Sun—the dark side—is 250°F below zero.

The side of the Earth that is closest to the Moon is subject to a greater tug of gravity than the center of the Earth. The side of the Earth facing away from the Moon experiences less of a tug than the center of the Earth. This has the effect of stretching the Earth a bit, and the parts that stick out are referred to as "tidal bulges." The most noticeable effect is the tides, but the Earth itself is made slightly oblong by a few inches. This tidal bulge is in addition to the bulge created by the Earth's rotation.

The Earth rotates once every twenty-four hours, faster than the Moon orbits it, which is about twenty-nine days. That tidal bulge wants to speed up the Moon and nudge it ahead in its orbit. The Moon resists by pulling back on the tidal bulge. That slows the Earth's rotation. Because of conservation of energy, whatever one body loses, another must gain. The Moon gains by having a bigger orbit, which means moving slightly away from the Earth. The Moon is moving away from the Earth about an inch and a half per year.

87. What is the big red spot on Jupiter?

The big red spot, officially known as the Great Red Spot, is a huge, long-lasting storm in the atmosphere in the southern hemisphere of Jupiter. It is about seventeen thousand miles across and is so big that three Earths would fit inside. It is a high-pressure storm, much like a giant hurricane, and it's the biggest storm in our solar system. Hurricanes on Earth are low-pressure systems, but Jupiter's red spot is a high-pressure system that rotates counter-clockwise with a period (time for one rotation) of about seven days. The Great Red Spot's big red spot is actually pink and orange and was discovered in 1664, when Robert Hooke, English scientist and architect, saw it in his telescope. Italian-French astronomer Giovanni Cassini is credited with the discovery of the Great Red Spot at about the same time.

The colors are the result of chemical reactions occurring in the atmosphere. The best theory is that the storm dredges up material, mainly phosphorus, from the interior of Jupiter. That giant red spot has been raging for over three hundred and forty-five years.

Jupiter is a huge gas ball. It does not have a clear boundary between atmosphere and solid surface like Earth. The gases get thicker and thicker the closer they get to the center and most likely, at some point, change to liquid. There may even be a solid core at the center of Jupiter. Scientists simply don't know for sure.

Jupiter is 2.5 times as massive as all the other planets combined, which is probably why it is named after Jupiter, the king of the Roman gods. Jupiter is 318 times as massive as the Earth, it has sixty-seven moons that we know of, and it is encircled by a ring system, just like Saturn. The planet's four largest moons are called the Galilean moons, named after Galileo, who discovered them and plotted their positions in 1610. The largest, Ganymede, is bigger than the planet Mercury. You can see Jupiter and its four largest moons with a small telescope or a good pair of binoculars. However, the rings of Jupiter are too nebulous and flimsy to see from Earth.

A Jupiter day is about ten hours, the time it takes for one complete rotation. It takes Jupiter about twelve of our years to go around the Sun once. Jupiter is composed almost entirely of hydrogen. If Jupiter were bigger, its gravitational pull would be strong enough to crush this material into a small core, which would trigger nuclear fusion and turn Jupiter into a star, much like our Sun.

88. Why does the Moon change color and size?

Two very good questions. First, let's deal with the size. Yes, the Moon does appear to be larger when it is on the horizon and smaller when overhead (see page 81). This is a result of the Ponzo illusion, named after Italian psychologist Mario Ponzo, who published his findings in 1913.

The brain interprets the sky as being farther away near the horizon and closer near the zenith, or directly overhead. We notice this on a cloudy day, when the overhead clouds may be a few thousand feet away, but near the horizon they might be hundreds of miles away. Since the Moon actually stays the same size, our brain makes it look bigger when it's near the horizon to compensate for the increased distance. The Moon on the horizon is interpreted by the brain as being farther away. Because it is the same apparent size as when it's high up, the brain figures it must be physically larger. Otherwise, the distance would make it look smaller.

Try this fun activity: Hold the tip of a pencil eraser at arm's length and aim it at the rising Moon. Note the size of the eraser and the size of the Moon. Try this again when the Moon is more overhead. You'll find the relative sizes haven't changed at all.

Now for the color question. Quite often, the Moon will appear yellowish, orange-colored, or reddish, especially when we have a full moon that is rising in the eastern sky in the early evening. Later in the evening, when the Moon is higher or more overhead in the heavens, the same Moon looks white or light blue in color. The reason for this change of color rests in the nature of light itself.

Remember that mnemonic device for learning the colors of the rainbow, ROY G BIV, for red, orange, yellow, green, blue, indigo, and violet? The ROY colors (red, orange, yellow) have the longest waves, and the BIV colors (blue, indigo, violet) have the shortest. We can think of light in terms of waves we see on water. Visible light is electromagnetic radiation that has the properties of frequency, wavelength, and velocity. Frequency is the number of vibrations per second. Wavelength is the distance from the crest of one wave to the crest of the next wave. Velocity is the speed of a wave, or how fast the wave travels. Light waves, unlike water waves, do not need a medium to travel through.

In a phenomenon known as scattering, particles in the atmosphere, such as smog, dust, moisture, and even oxygen, act as tiny mirrors that reflect light in every direction. The Moon looks orange or reddish due to this scattering of light by the atmosphere. When the Moon is on the horizon, the moonlight has to pass through much more atmosphere as it travels to our eyes than when the Moon is overhead. By the time the light from the Moon reaches our eyes, most of the blue and green wavelengths have been scattered. What is left is the red, orange, and yellow part of the spectrum. The same phenomena is responsible for our red sunrises and red sunsets.

The particles that reflect light are very tiny, typically one-tenth the length of a wave of visible light. The shorter the wave of light, the more the scattering. A red wave is too long to effectively reflect off a particle that is a fraction of the length of that wave. We can think in terms of the "mirror" particle being too short. But a blue wave is short enough to bounce off the same particle. So blue light can't travel through very much atmosphere before it is reflected in every direction. That's why the BIV waves have all dissipated by the time the moonlight reaches our eye. Only the ROY colors remain, so we see the moon as reddish or orange.

When that same Moon is overhead a few hours later, the moonlight does not have as much atmosphere to travel through, so all the colors get through, and we see the Moon as more whitish or bluish.

89. How do clouds form, and how do they get their color?

Clouds are bunches of very tiny water droplets. The droplets are so light and small that they can float in the air. Clouds form when warm air rises, expands, and cools. The warm air has some moisture in it, but it is in vapor form and cannot be seen. The rising air cools about 3.5°F for every thousand feet in altitude, a pace referred to as the adiabatic cooling rate.

The lifted air is cooled to its dew-point temperature, the temperature at which air is completed saturated and can no longer retain its vapor. The relative humidity has reached 100 percent.

The expelled vapor condenses onto dust particles and oxygen molecules. Most of these particles come from cars, trucks, volcanoes, and forest fires. To "condense" means to change from a gas state to a liquid state. Billions and billions of these tiny droplets come together and form what we see as a visible cloud. At the highest altitudes the tiny droplets will freeze and form ice crystals.

So why are clouds white? Clouds are white because they reflect light from the Sun, which is made of the seven colors of red, orange, yellow, green, blue, indigo, and violet. Most of the time, clouds reflect those seven colors evenly, in about the same amount, to give white light. We see grayish, or very dark, clouds if the cloud cover is very thick; not as much sunlight is getting through the clouds. Thunderstorm clouds go so high in the atmosphere that these clouds take on a blackish appearance. The sheer thickness of the cloud layer blocks a greater amount of sunlight, and very little light reaches our eye, hence the "blackish" tint of the thundercloud.

Cloud height is determined by the type of cloud and the size of the tiny water droplets that make it up. Those fluffy white cumulus clouds with the flat bottoms generally have the heftiest water droplets, so those clouds are generally found below six thousand feet, but in some cases they can be found much higher. Cumulus clouds move about 10 to 20 mph but faster with thunderstorms.

Those very high thin wispy clouds, often called mare's tails, are composed of very small ice crystals and are spotted at higher altitudes, beginning at about thirty thousand feet above the ground. These high cirrus clouds are pushed by the jet stream and can reach speeds of over 100 mph. The presence of a lot of cirrus clouds indicates that a change in weather is fast approaching.

Fog is a special kind of cloud. Most fog forms when warm moisture-laden air flows over a colder surface. If the air is full of moisture, it will condense onto particles in the air.

Besides the two main types of clouds, cumulus and cirrus, there are stratus clouds. "Stratus" comes from the Latin for "to spread out"; these stratus clouds stretch out over great distances in the sky. Stratus clouds yield a long, steady rain.

As a general rule, we tend to see many cumulus clouds in summer and a preponderance of stratus clouds in winter.

90. Early in the morning, I saw a bright light moving across the sky. What was it?

You most likely saw the International Space Station go over. But there are several other possibilities: shooting star, comet, airplane, or satellite.

What about a shooting star, also called a falling star or meteor? That's a very fast streak of light that lasts a few seconds at most. The streak appears suddenly, goes only part of the way across the sky, and disappears. Not likely. It could also have been an airplane. They move slowly across the night sky, depending on altitude. Many times you can spot their red beacon light and even the position lights on either wing. Still, a satellite is most likely, because they move slowly. Many are going in a north-south direction, especially the reconnaissance, weather, and Earth-resources-sensing satellites.

The satellites used for television reception, like the ones used by DirecTV and Dish Network, have geosynchronous orbits, which means they stay over the same location on the Earth (see page 140). Their period, or time to make one revolution around the Earth, is twenty-four hours. You will recognize this as the time for the Earth itself to rotate once on its axis. They orbit at 22,000 miles above the Earth, which synchronizes their orbit to the rotation rate of the Earth. These satellites always follow along an extension of the equator. You wouldn't see these satellites moving across the sky.

So most likely what you spotted was the International Space Station (ISS) traveling from west to east. The ISS has grown very big over the past few years. They've added module after module and solar panels galore. Its altitude is about 250 miles above the surface of the Earth, moving about 17,000 mph and taking about 1.5 hours to make one orbit of the Earth. There are other satellites in orbit around the Earth, but none as bright as the International Space Station.

You can see the ISS either in the early-morning hours before dawn or a few hours after sunset. It may be close to dark for the observer on Earth, but the sunlight strikes the surfaces of the ISS and reflects down to us. For an observer on the ground, it takes about 1.5 minutes for the ISS to go from horizon to horizon.

A really neat website for any sky observer is heavens-above.com. Set up a simple password and put in the coordinates of your house. The website will give a chart of when the ISS passes over your home, the time when it comes above the western horizon, when it reaches its highest point, and when it goes down below the eastern horizon. Also, all the observable passes of the Hubble Space Telescope are given.

Here is something for the curious mind and adventurous soul! Try to spot an Iridium flare. Sixty-six Iridium Communications satellites are up there. They're not very big, only about thirteen feet long and three feet wide. Each satellite has three flat, door-size, highly polished aluminum surfaces (combo solar panel and antenna) that can reflect sunlight just like a mirror. The axis of the satellite is maintained vertical to the Earth's surface. So software can predict the date, time, and any position on Earth where the bright light from the Sun hits those antennas and reflects to Earth. What we see here on Earth is a light or short streak of light, getting brighter until it reaches maximum brightness and fades again to nothing. The whole thing lasts only about five to twenty seconds.

The heavens-above.com website will tell you where to look in the sky, the exact altitude, and the azimuth (direction from north). The time of day will be right before sunrise or right after sunset, when the Sun is below the horizon for the observer but the satellite is high enough that the Sun's rays strike it. People often report that brief flash of light as a UFO. Happy satellite hunting!

91. Is it possible for a human to go to Mars?

Yes, we can go to Mars. But the journey would be expensive, dangerous, and long. It doesn't help that there don't seem to be any strong arguments to go to Mars for economic reasons, either. For example, Mars has a lot of iron in its soil, but we have plenty of iron here on Earth.

We won't go to Mars to get there first. That's what the Moon race was all about in the 1960s, during the Cold War. We just had to beat the Russians to the Moon to prove our technological superiority. Yes, there was a lot of science that came out of the Moon race, and a lot of products, and processes, and spin-offs. But the Moon race was more political than scientific.

While it takes only three days to get to the Moon, a trip to Mars takes eight months one way. You'll want to stay for a while, and it's another eight months back, so a round trip to Mars and back would probably take at least two years. It's going to take a very special person to handle two years in a confined and sterile environment, far from family and friends, with no access to medical facilities, accompanied by the same people every day.

You can't go to Mars any old time you feel like it. A favorable period for sending a spacecraft to Mars is called a launch window. A launch window is a specific time period when any rocket or spacecraft must lift off to achieve a particular landing, mission, or rendezvous with another body. The launch window time is determined by the positions in their orbits of the launching platform, which is usually the Earth and the target, namely Mars. A launch window for Mars opens every twenty-six months. Launching a spacecraft to Mars is similar to throwing a dart at a moving target, except that you, the thrower, are standing on a rotating platform.

The spacecraft, launched from Earth, would actually go into an orbit around the Sun. The craft would then intercept the orbit of Mars about eight months after launch. Entry into the Martian atmosphere is tricky. Numerous unmanned missions have failed during this crucial stage, because so many things can go wrong: electronic glitches, lack of fuel, spacecraft speed, sandstorms, faulty trajectory, and rock outcroppings.

Another big problem is the communications lag between Earth and Mars. The radio time between the Earth and the Moon is only about 1.5 seconds. It can take up to about twenty minutes to send a radio call to Mars, and another twenty to get a response. So if the astronauts make a call to NASA for help or with a question, it could be about forty minutes before they get an answer, and that's if NASA answers as soon as it gets the message.

Mars is no place to spend vacation time. It is a hostile environment, with an atmosphere that is almost all carbon dioxide and so thin it holds no heat. Average temperature at the equator is −50°F. The atmospheric pressure is too low and the temperature too cold for regular liquid water to exist.

My guess is that we will continue to explore Mars with unmanned spacecraft for many years to come. Since 1960, according to the Planetary Society, almost forty craft have been sent to Mars. The *Mars Reconnaissance Orbiter*, sent in August 2005, mapped Mars from low orbit. The *Mars Phoenix Mars Lander*, launched in 2007, touched down on Mars in May 2008. It landed near one of the polar ice caps and dug down 1.5 feet searching for ice and water.

In 2011, NASA sent a sophisticated Mars Science Lab rover vehicle to the Red Planet. On August 6, 2012, NASA successfully landed *Curiosity* in Gale Crater on the surface of Mars. NASA asked the general public to name the spacecraft. Its website received hundreds of possible names, including Adventure, Pursuit, Vision, Wonder, and Perception. A sixth-grader in Kansas submitted the winning entry, Curiosity. Clara Ma wrote, "Curiosity is the passion that drives us through our everyday lives. We have become explorers and scientists with our need to ask questions and to wonder."

92. What are Saturn's rings made of?

If you ever get an opportunity to peer at Saturn through a telescope, even a small one, you will be looking at one of nature's most majestic sights. You will be inspired, as Galileo was in 1610, when he became the first person to observe the beautiful ring structure of Saturn.

Four robot spacecraft have visited Saturn. The latest, named *Cassini*, went into orbit around Saturn in July 2004. *Cassini* sent back thousands of color pictures of the rings and moons of Saturn, and it carried a detachable vehicle, *Huygens*, which parachuted onto the surface of Titan, a moon of Saturn. Stunning pictures of this alien world were sent back to Earth by the wheelbarrow-size *Huygens*.

There are between five hundred and one thousand rings, with gaps in the rings, and they are mainly composed of ice, along with some rock and dust. Collectively, they are 175,000 miles wide but only about three hundred feet thick. The smallest particles are smaller than a grain of sand, and the largest are about the size of a bus. Each little particle, no matter what its size, could be thought of as a moon.

Saturn has sixty-two moons, some major and some minor. Some of the minor moons are only a mile or two across. Titan is the largest moon and is much bigger than our own Moon.

We know that the Moon pulls on the Earth and the Earth pulls on the Moon. It is that gravitational tug on each other that causes the tides here on Earth (see page 129). But tidal forces affect solid objects as well. When an object, such as a Moon, gets too close to the planet it orbits, the tidal forces will tear that object apart and shatter it into thousands of pieces. A mathematical rule, known as the Roche limit, determines how close an object can get to a planet without being torn apart. The Earth's Roche limit is about 12,000 miles. Not to worry, though; our Moon is 240,000 miles away.

Saturn is almost one hundred times more massive than our Earth, so its Roche limit extends out quite far from the planet's surface. Billions of years ago, moons, rocks, comets, and any other debris that got too close to Saturn were pulverized and trapped in orbits, which formed the rings. Even today, asteroids and other objects are continually bombarding the solar system, and any of that stuff that comes too close to Saturn gets caught in the ring system.

The other gas giants, Jupiter, Uranus, and Neptune, also have rings around them, but their rings are quite faint and not easily seen. They are more like wispy circles. Saturn has the most material inside that Roche limit, and hence has the most elaborate set of rings.

There are many gaps in the rings. The biggest gap is the Cassini Division. From Earth, it looks like a thin black gap in the rings. It was first seen in 1675 by Giovanni Cassini from the Paris Observatory.

93. How close can you get to the Sun without burning up?

Not very close. The temperature on the surface of the Sun is about 10,000°F. Compare that with your typical commercial pizza oven of around 700°F.

The Sun is 93,000,000 miles from the Earth. If you drove a car at 65 mph, without stopping, it would take 160 years to get there. The Sun is 400 times farther away from us than the Moon.

If you used an aluminum spacecraft, you could get to within 8 million miles of the Sun. Proportionally, that's equivalent to being on the nine-yard line in football. The temperature would be about 1,220°F, the melting temperature of aluminum. The trick to getting close to the Sun is to radiate or reflect heat as fast as it is absorbed. One way to do this is to paint the side of the spacecraft toward the Sun silver or white. Paint the side away from the Sun a black color. White reflects and black absorbs. Another device is to make the spacecraft long and thin, like a needle, and point the "needle" toward the Sun. That technique will dump more heat. Less surface area of your craft is exposed to the Sun. You're up to the twelve-yard line . . . and sweating. A third mechanism is to coat your spaceship with the kind of tile that was used on the space shuttle. That reinforced carbon-carbon heat shield will withstand temperatures of up to 4,700°. Now you can get to about 1.9 million miles from the Sun. You've made it to the two-yard line; you are in the red zone.

Getting any closer would be a hellish trip, no pun intended. Cosmic radiation on that two-yard line would kill passengers inside the spacecraft within hours.

The *Mercury Messenger* spacecraft was launched by NASA in 2004. It was a daring and complex mission, in which the refrigerator-size vehicle made one pass by Earth, two passes over Venus, and three passes by Mercury before going into orbit around Mercury in March 2011. *Mercury Messenger* came within 30 million miles of the Sun. That's not even to the fifty-yard line, to use our football analogy. *Mercury Messenger* employed a large ceramic cloth sunshade. The *Mercury Messenger* spacecraft photographed the entire surface of Mercury, studied the soil composition, mapped the magnetic fields, and tested for any atmosphere. In November 2010, NASA announced the finding of water, ice, and organic compounds in permanently shadowed craters near Mercury's North Pole.

But for humans, getting really close to the Sun is not practical or wise. We can just lie out in the backyard and get a real nice sunburn in less than an hour.

94. How long would it take to travel from Earth to Venus in a rocket ship?

It takes about five months, or 150 days, to get to Venus, the planet named after the Roman goddess of love and beauty. A straight, direct shot from Earth to Venus is not the way to go. It would require too much propulsion power and rocket fuel. No country has such a powerful rocket. There is a much easier way. A spacecraft is put into a path, or orbit, around the Sun so that at some later time, the orbit of the spacecraft intersects the orbit of Venus.

Space navigators use Newton's laws, Kepler's laws, and a Hohmann transfer technique to get spacecraft to Venus and Mercury and to the planets of Mars, Jupiter, Saturn, Uranus, and Neptune.

Newton's three laws of motion are the foundation of mechanics in physics. The first law is the law of inertia, the second is the law of acceleration (force equals mass times acceleration, or $F = ma$), and the third is the law of action and reaction; all are employed in determining the forces, paths, and techniques used in space travel.

Johannes Kepler developed three laws that describe the motion of the planets across the sky. The first and third are vital to space travel. The first is the law of orbits: All planets move in elliptical orbits, with the Sun at one focus. The third law relates the time it takes to complete an orbit versus the distance away from the Sun.

A German engineer and rocket enthusiast, Walter Hohmann, wrote a book in 1925 detailing how a spacecraft can transfer from one orbit to another by using two engine burns at precisely the right time, direction, amount of thrust, and duration.

You can't go to Venus just any old time. A window of opportunity presents itself every nineteen months. That is when Venus is fifty-four degrees behind the Earth in its orbit. The time of day for launch is also important. To go to Venus, the spacecraft must be launched within a few minutes of 7:40 AM local time. This time places the spacecraft on the launch pad in the precise position (Sun off to the tangent east) to make the initial thrust to put the craft on its way to Venus (see page 120).

Despite the obstacles, nearly thirty spacecraft missions have been sent to Venus by the United States, Russia, and the Europeans. The United States'

Magellan spacecraft used radar to map over 98 percent of the surface of Venus. It started its mission in 1990. The European Space Agency (ESA) launched the *Venus Express* in November 2005, and it went into orbit around Venus in April 2006. The *Venus Express* is loaded with scientific instruments and successfully photographed lightning in the clouds of Venus. Japan launched the spacecraft *Akatsuki* in 2010, and it failed to go into orbit around Venus. *Akatsuki* is in orbit around the Sun. The spacecraft will be in the correct position in its orbit in 2015 to make another orbit-insertion burn.

Venus is the second planet from the Sun. Because its size, density, gravity, and composition are similar to those of Earth, Venus is sometimes referred to as our "sister planet." But it is a mean sister. The temperature on the surface is hotter than a pizza oven's, about 900°F.

Venus is much like a car parked out in the Sun on a summer day. Sunlight comes through the windows and is absorbed by the materials in the interior. The seats, dashboard, etc., radiate that heat as infrared light. Infrared waves are a lot longer than visible light rays, and infrared light can't go through the window glass. So the heat is trapped in the car and the temperature goes way up. The thick carbon dioxide and the sulfuric acid clouds of Venus are like the glass in the car window. Both represent the runaway greenhouse effect (see page 271).

The air pressure on Venus is ninety times that on Earth. Such high pressure crushed, for example, Russian spacecraft that had landed on the planet. The Russians sent ten probes to the surface of Venus, and while all could be called successful to some extent, the longest time that any Russian lander survived was just over two hours.

Venus's carbon dioxide atmosphere is denser than the air that you and I breathe. Surface conditions on Venus are horrific. The surface is bone-dry, and the heat of 900°F causes water to boil away. Much of the surface is barren plains punctuated by thousands of volcanoes and four large mountain ranges.

Venus reflects much sunlight off its clouds, so it shows up in the night sky in the west or in the morning sky in the east. It is so bright that gunners in World War II were known to have shot at it, thinking it was an enemy aircraft. Venus is often reported as a UFO.

Venus takes 243 days to rotate on its axis and 225 days to go around the Sun. Hard to believe, but its day is longer than its year. While all the other planets rotate counterclockwise as viewed from above the solar system, Venus rotates clockwise.

95. If the Moon is so heavy, why doesn't it fall?

The Moon actually does fall. Strange as it may seem, the Moon is falling all the way around the Earth as it orbits our planet. But to get a more complete explanation with deeper understanding, we must look at the concept of forces.

A force is a push or a pull. One of the most common forces is gravity. We know that gravity acts on an object on Earth by pulling it straight down toward the center of the Earth.

However, just because there might be a force on an object, this doesn't mean that the object will go in the direction of the force. Pretend a bowling ball is moving straight down the middle of the lane. You run up to the bowling ball and give it a sideways kick toward the gutter. The bowling ball does not go straight into the gutter and does not continue in the middle-of-the-lane path it originally had. Instead, the ball changes direction slightly so it has a diagonal motion, and it continues rolling at an angle. This is because the ball already had some forward motion to it before you kicked it, so the force of your kick combined with the preexisting force somewhat changes the ball's direction.

Now pretend you drop a baseball from the rim of a tall cliff. It will fall straight down. The only force acting on it is gravity. If you throw the baseball straight out, horizontally, it will move horizontally, but at the same time, it will start to fall some. Remember, gravity is still pulling it down. The baseball falls at an angle that constantly changes as the force of the throw diminishes. The path of the ball is an arc, actually called a parabola.

Next time, you throw the baseball harder. It goes farther, falling at a shallower, more gradual, angle. The force of gravity was the same, but your throw gave the baseball greater forward speed, or velocity, so the deflection is much less, resulting in a shallower angle.

If you threw that ball so hard that it traveled about one mile before it hit the ground, it would have to fall about six inches more than before. Why is that? The Earth is curved. So as the ball is traveling out that one mile, the Earth is curving away underneath it.

Now you throw the baseball even harder, perhaps hard enough to make a major league roster! If you throw the ball straight out, so it travels six miles, the Earth would curve away about thirty feet. Throw it sixty miles out, and by the time the baseball drops, the Earth will have curved more than a half mile away.

Finally, you ate your spinach, you feel like Superman, and you throw the baseball so hard that the Earth curves away underneath the ball so much that the ball never gets any closer to the ground. It goes all the way around in a circle and might hit you in the back of the head. You have just put the baseball into orbit around the Earth.

In actual practice, you couldn't do this at the surface of the Earth, because the baseball would run into too much air resistance. You'd have to get it up quite high, about 100 miles, before you threw that baseball straight out. This is the mechanism we use to put satellites around the Earth. Our natural satellite, the Moon, stays in orbit for the same reason. All objects, in the absence of air resistance, fall at the same rate, so the mass or weight of the satellite does not make any difference. For Earth, the speed you need to throw a baseball or satellite so that the curve of its drop, or fall, matches the curvature of the Earth is five miles a second, or 18,000 mph.

Gravity gets weaker the farther out you go, which is why the Earth's gravity has a much weaker pull on the Moon than it does on Earth-observation satellites, the International Space Station, and Hubble Space Telescope. Those satellites are about 200 to 500 miles above the surface of the Earth, whereas the Moon is roughly 240,000 miles from the Earth. The Moon orbits the Earth much more slowly than those low-Earth-orbit satellites. The speed of the Moon in its orbit around the Earth is about 2,300 mph. It takes a full month for the Moon to make an orbit around the Earth. Those artificial satellites go around the Earth at a speed of around 17,500 mph, taking only ninety minutes to orbit the Earth.

Here is a special case. If a satellite is put up about 22,500 miles from the Earth, it takes twenty-four hours to orbit the Earth—the same time it takes the Earth to rotate, which means the satellite's following a geosynchronous orbit, so it stays above the same spot on the Earth (see page 140). So we can use a small dish antenna and pick up those TV channels without having to move the antenna.

96. How big is our galaxy?

Very big! The Milky Way Galaxy is one of billions of galaxies in the known universe. The Milky Way is a barred spiral galaxy—a galaxy with a central bar of bright stars across its center—in what is called the Local Group within the Virgo Supercluster of galaxies. It is estimated that our Milky Way Galaxy contains about 200 billion to 400 billion stars.

The disk of the Milky Way Galaxy is about one hundred thousand light-years in diameter. A light-year is a unit of distance, not a unit of time. It is the distance that light travels in one year. Light travels at about 186,000 miles a second. So multiply the number of seconds in a year, roughly 31,500,000, by 186,000 miles per second, and a light-year comes out to be about 6 trillion miles. That is 6 million million miles, or, written out, 6,000,000,000,000 miles. And the diameter of our galaxy is about one hundred thousand of these!

Our solar system, with the Sun as the primary star, is located about one-third of the way out in one of the spiral arms of the Milky Way Galaxy. If the entire galaxy were reduced to eighty miles in diameter, our solar system would be only 0.1 inch in diameter. We are indeed small stuff in a big place!

Go out in the country, away from the city lights, look up, and note the hazy band of white light running from the southwest to the northeast. You are looking edgewise at the heart of the Milky Way. The Milky Way Galaxy appears brightest in the direction of Sagittarius, which is toward the Galaxy's center.

Be aware that every star you see with the naked eye in the night sky is in our own Milky Way Galaxy, with one exception. The Andromeda Galaxy in the constellation Andromeda looks like a white, fuzzy patch of cotton candy. The galaxy, also known as M31, can be seen right off the square of stars in the constellation Pegasus. Use any star chart or the Internet to find its position.

The Andromeda Galaxy is one of about fifty galaxies in the Local Group and is a spiral galaxy. It is about three million light-years away. When you look up at the Andromeda Galaxy, the light you see tonight left the galaxy three million years ago, which makes looking at the night sky, in one sense, like being in a time machine (see page 339).

The Milky Way's center is home to a very dense object believed to be a supermassive black hole. Most every observed galaxy has one or possibly more.

There is a television ad that says the universe has more stars in it than there are grains of sand on the Earth. According to the best estimates, there are one hundred stars for every grain of sand on Earth.

97. How does the Moon affect the ocean tides?

The gravitational pull of the Moon affects tides on the Earth. As the Moon's gravity pulls on the water of the Earth, the ocean bulges on both sides of the planet. The Earth's water bulges on its Moon-facing side because it's being pulled by the Moon's gravity, leading to a rise in the tide. But the Moon's gravity also creates a high tide on the opposite side of Earth, because the Moon is pulling on the Earth, too. The inertia of the ocean water causes it to remain at rest, creating this second bulge as the Earth moves away from its water. The solid Earth is being pulled toward the Moon, but liquid water on the far side does not have as much pull or tug.

Points on the sides of the Earth between the two tide bulges have a low tide. Because the Earth takes twenty-four hours to rotate once, any point on Earth will have a high tide followed by a low tide in roughly six-hour intervals. During one twenty-four-hour period, each position on Earth will have about two high tides and two low tides. Because the Moon is moving around the Earth, the actual time between tides is about six hours and thirteen minutes. If the Moon were somehow stopped in its orbit, the time between tides would be about six hours. But during one rotation of the Earth, the Moon has moved ahead into its orbit about one-thirtieth of the distance of one orbit, or about twelve degrees. So the Moon is not in the same position in its orbit as it was twenty-four hours ago.

The Sun affects ocean tides, too. The Sun's gravity is much stronger than the Moon's, but the Sun is much farther away from the Earth, so it has a weaker effect on the tide. The Moon's gravity causes about 56 percent of the tide, and the Sun's gravity about half the Moon's effect.

When the Sun, Moon, and Earth are all lined up, gravity's pull is the greatest. These tides occur twice a month, during the full moon and the new moon. They are called spring tides. Despite the name, they have nothing to do with the season.

When the Moon is in its first quarter and its third quarter, the Moon and the Sun are at ninety degrees from each other, so the gravitational pull of one counteracts the other's. The lower-than-normal tides that result are termed neap tides.

Tides are also more pronounced once a year when both the Sun and the Moon are closer to the Earth. This happens around January 4 of

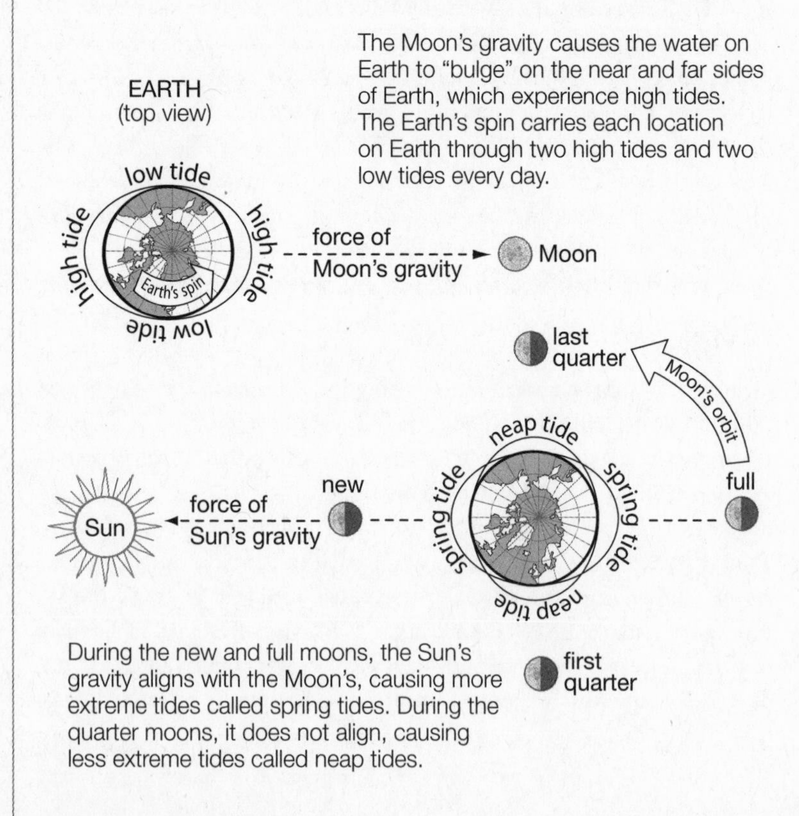

every year. The Nazis fire-bombed London on December 29, 1940, knowing that there would be little water in the Thames River at low tide, making London's fire hoses ineffective. (Yes, hoses; they couldn't rely on hydrants, which required a pumping station, so they literally carried the fire hoses through the mud and placed them in the river, relying on fire-engine-driven pumps to pull the water out of the river.)

The famous beach assaults in World War II, such as in Italy and Normandy and against the Japanese in the Pacific, were planned for high tide. The idea was to get men and material as far inland as possible. Arguably the most difficult and brilliant amphibious landings were designed by General Douglas MacArthur to outflank North Korean troops that had trapped UN forces in the Pusan perimeter. The Inchon landings on September 15, 1950, were tricky because of the tides, seawall, mud flats, and monsoons.

Tides are most pronounced along the coastline and in bays, where there is a unique topography. The tides at the Bay of Fundy between Nova Scotia and New Brunswick have a range of fifty feet between high tide and low tide. An understanding of tides is important for the fishing industry, for coastal navigation, and for the building industry along the coast.

98. When is Halley's Comet going to come around again?

Halley's Comet is returning in July 2061, about fifty years from now. Edmond Halley was an English astronomer and a friend to the most famous scientist of all time, Isaac Newton. In 1705, he used Newton's law of gravitation to find the orbits of comets from their known positions in the sky over a period of hundreds of years. Halley determined that the comets sighted in 1531, 1607, and 1682 had almost the same orbit. He used gravitational calculations to predict the return of this same comet in 1758. Indeed, the comet was sighted on Christmas night of 1758, but Halley did not live to see his prediction come true. He had died some sixteen years earlier, in 1742. Halley is honored by having the comet named after him. The last time Halley's Comet "came around" was 1986. Because of its orbit and position relative to the Earth, Halley did not put on a good show for anyone in North America.

Going back through the records, it has been shown that Halley's Comet has been sighted almost every seventy-six years as far back as 240 BC. The closest sightings at the time of Jesus were 12 BC and again in 66 AD. So Jesus never saw Halley's Comet.

When Halley's Comet was seen in England early in 1066, it was viewed as a bad omen. Later that year, William the Conqueror invaded England and defeated Harold II at the Battle of Hastings. Halley's Comet is prominently displayed in the famous Bayeux Tapestry that commemorates the Norman king William's victory over the Saxon king Harold.

Mark Twain said, "I came in with Halley's Comet in 1835. It's coming again next year [1910], and I expect to go out with it." And he did.

A comet has been described as a "dirty snowball." The nucleus is composed of ice, dust, and frozen methane. The nucleus of Halley's Comet is potato-shaped, ten miles long and six miles across. When it comes close to the Sun, it develops a temporary atmosphere called a coma (from the Greek word for "hair"), a dense cloud of water and carbon dioxide that surrounds the nucleus. Two long tails, one a dust tail and the other an ion tail, stretch back millions of miles. The tail of dust is created by the comet material being burned off the giant snowball. The dust tail forms a curved path that streams behind the comet in its orbit around the Sun. The ion tail points directly away from the Sun following the magnetic lines of force that emanate from the Sun.

Most comets have cigar-shaped orbits and can only be seen with the naked eye when they are close to the Earth and Sun. More than four thousand comets have been cataloged. Meteor showers occur when the Earth passes through the debris left in the path of a comet. The Perseid meteor shower occurs every year between August 9 and 13, when the Earth passes through the orbit of Comet Swift-Tuttle. Halley's Comet is the source of the Orionid meteor shower every October.

99. How do planets move?

Once the planets were in motion, nothing was needed to keep them in motion. The best current thinking is that our solar system—the Sun and eight planets we know (nine if you include Pluto, which is no longer officially a planet)—was formed out of a swirling disc of gas and dust some four to five billion years ago. The main body of this mass heated up to a temperature at which nuclear fusion could start. That became the Sun. The outer, smaller clumps of dust and gas in this huge disc formed the planets. Scientists believe that other star systems are formed in much the same way.

So once this rotation is in motion, no force or push or pull is needed to keep it rotating. That neat idea follows Isaac Newton's universal first law of motion, the law of inertia: An object in motion tends to stay in motion, and an object at rest tends to stay at rest.

The planets move in circles around the Sun. More precisely, the orbits are ellipses, or squished circles, if you will. Venus is the planet with the most circular orbit, and Mercury has the most elongated, or egg-shaped, orbit, if we exclude Pluto.

It is generally recognized that life needs both a source of energy and a solvent. For us Earthlings, the source of energy is the Sun and the solvent is water. In order for life to thrive, the planet must lie within a narrow range of distances from the star or sun it is going around. If the planet is too close to the star, the solvent (water) will boil away. If it is too far away from the sun, the solvent will freeze. For our Sun, the habitable range seems to be between the orbits of Venus and Mars. It just so happens that Earth is nearly halfway between these planets. We hit the jackpot!

Mercury is too close to the Sun for people to live. Any water would evaporate immediately. Jupiter is too far away, and it's a gaseous planet, so it doesn't have a solid surface anyway. Mars is not the nicest place to live; it's too cold most of the time. Venus has dense carbon dioxide clouds dripping sulfuric acid and is about as hot as a pizza oven. Not a nice place to visit! We humans are very fortunate to have this "pale blue dot" to live on.

100. What are the chances of an asteroid hitting the United States?

Short answer: possible, but not probable. However, asteroid, meteor, and comet impacts have to be taken seriously. Natural disasters, such as the recent Indian Ocean tsunami, can kill tens or hundreds of thousands of people. But an asteroid, meteor, or comet collision with Earth could kill millions or even billions of people. A loss of a million people is a terrible disaster; a billion, or a thousand millions, is on another order of magnitude altogether.

The February 15, 2013, meteor blast in the Ural Mountain region of Russia, near the city of Chelyabinsk, just might have been a wake-up call to the power of an exploding meteor. More than a thousand people were injured and 7,200 houses damaged from the shock wave of the blast. The meteorite was estimated to be more than fifty feet in diameter and have a weight about 11,000 tons.

A recent warning from NASA was issued on December 25, 2004. Looking ahead, an asteroid called 2004 MN4, with a diameter of 1,350 feet, has a one in three hundred chance of hitting Earth on April 13, 2029. Astronomers keep tracking some seven hundred of these "near-Earth asteroids." NASA stresses that the risk probably will be reduced to zero as they "keep an eye" on its orbit.

Two fine movies, *Deep Impact* and *Armageddon*, both released in 1998, deal with the subject of comets or asteroids striking the Earth. In both movies, crews in spacecraft are sent to intercept and destroy the invader. If the time ever comes for us to actually need to do something like this, the hope is that we Earthlings will send unmanned devices to do the job.

Scientists have identified more than 200 proven sites of meteor impact on planet Earth. Let's look at three of the best known.

A comet or asteroid five to ten miles wide struck in the Yucatán Peninsula in Mexico 65 million years ago. This impact is believed to have caused the extinction of the dinosaurs and thousands of plant and animal species. The best proof came when scientists Luis and Walter Alvarez discovered a thin layer of iridium in soil that covers the entire Earth. The metal iridium is rare on Earth but abundant in meteorites.

A second example of meteor impact is the Barringer Crater, which was created fifty thousand years ago when a nickel-iron meteorite 150 feet across slammed into the desert floor in Arizona. The crater is about three-fourths of a mile wide and 750 feet deep. When Europeans first came across the crater, they discovered thirty tons of meteoritic iron scattered around the crater. The force generated by the impact was equal to the explosion of 2.5 million tons of TNT.

In the third instance, a large meteor or piece of an asteroid or comet exploded in the atmosphere in the Tunguska Region of Siberia in Russia on June 30, 1908. The blast devastated an area the size of Rhode Island.

On January 12, 2005, NASA launched the *Deep Impact* spacecraft, which arrived at comet Tempel 1 on July 4, 2005. The satellite probe smashed into the comet with a force of 4.5 tons of TNT and blasted a crater in it, which allowed scientists to study the composition of this space visitor. NASA scientists determined the ingredients that make up our solar system's primordial "soup."

So we're far along on having the technology to nudge a comet or asteroid into a different orbit and away from the Earth, and it's possible that we might one day soon be able to blast it with a nuclear bomb or deflect it with lasers.

In summary, an asteroid or comet hitting the Earth? Put this way down on the bottom of your list of things to worry about!

101. How many constellations are there, and how are constellations and stars named?

Constellations are groupings of stars that help us make sense of which stars are which. When you look at the night sky, you can count about fifteen hundred stars with the naked eye. Constellations break up the night sky into manageable bits. They help make sense of all those pinpoints of light.

Constellations were named thousands of years ago by different cultures in different lands. Wonderful mythologies and numerous stories evolved around the characters they saw in the starry heavens. The constellations we are familiar with today come from the ancient Babylonian, Greek, and Roman civilizations. There are eighty-eight of them in the night sky. The official boundaries of these constellations were finally established in 1929 by the International Astronomical Union (IAU), the organization of professional astronomers. Every star in the night sky is now in a defined constellation.

In my hometown in Wisconsin, I can see about fifty-three constellations over the course of the year. If you want to see all eighty-eight, you have to live near the equator. At the North Pole, you can see only about half of them in a year's time.

People in the northern part of the United States can see five major constellations on any night of the year: Ursa Major (the Big Dipper, or Great Bear), Ursa Minor (the Little Dipper, or Little Bear), Draco (the Dragon), Cepheus (the King), and Cassiopeia (the Queen). We refer to these as circumpolar constellations, because they seem to circle Polaris, the Pole Star.

These five constellations are good starting points to learn where other stars and constellations are located. For example, start with the Big Dipper, a part of the larger constellation of Ursa Major. The two end stars in the bowl of the Big Dipper point to Polaris, which is the last star in the handle of Ursa Minor.

There is a whole slew of these memory aids to help stargazers find their way around the night sky. For more information, there are many good books in the library that have star charts. There are also excellent websites to investigate; one of the best is skymaps.com.

Certain constellations are seen during different seasons, so spotting the constellations in the night sky each year is like seeing old friends. For example, Orion (the Hunter), is a winter constellation. As the leaves and temperature fall, go outside early in the evening and look to the east. Up comes Orion chasing Taurus (the Bull) across the night sky.

The names of stars are a blend of Arabic, Greek, and Latin words. One of the first star catalogs was put together by the Egyptian astronomer Claudius Ptolemy, who lived under Roman rule. Ptolemy, in about 100 AD, compiled the names of stars the Greeks had used, along with some Latin words. The Arab world had some very sharp astronomers, so the text was translated into Arabic, then into Latin, and finally into English and other modern languages. The names we use today are derived from that Arabic, translated to English. New star names these days come from the IAU.

Sirius, in the constellation Canis Major, means "scorching" or "searing" in Greek. It is an appropriate name because Sirius, the Dog Star, is the brightest star you and I will see from Tomah.

Castor and Pollux are stars in the Gemini Constellation. Gemini means "twins," and Castor and Pollux were twin brothers. Castor is generally understood to have been a Greek warrior. Arcturus, in the constellation Bootes, means "bear watcher." Arcturus follows the Great Bear as it revolves around Polaris, the Pole Star. Our own star, the Sun, simply goes by the name "Sun." But sometimes it is referred to by its Latin name, Sol.

About fifteen hundred stars can be seen by the naked eye, meaning not using telescopes or binoculars. These stars were all given names way back in ancient times. Constellation names depend on the culture that labeled them. Ursa Major, which we call the Great Bear, is called the Plow, or Plough, in England. The heart of Ursa Major is the seven bright stars that make up the Big Dipper. These seven stars are known in Hindu astronomy as the Seven Great Sages. In Dutch, it is called the Saucepan. In Finnish, its label is the Salmon Net. The handle of the Big Dipper has three stars, and the bowl has four stars. The middle star of the handle is actually two stars that revolve around each other. They are a binary pair. The brighter of the two is Mizar, and the dimmer is Alcor. The ability to see or resolve these two stars, which appear very close together, was often quoted as a test of eyesight. It is in Arabic writings, English literature, and Norwegian prose. The Plains Indians of North America used the binary system as an eyesight test for their young ones.

Try looking for the double star in the handle of the Big Dipper. Choose a moonless clear night when the humidity is low. Take along a pair of binoculars. That's what I do . . . now!

102. If fire needs oxygen to burn, how can the Sun burn in space, where there's no oxygen?

Yes, the Sun is a big ball of fire, but quite a different kind of fire from a match, fireplace, or bonfire. Instead, it is a nuclear furnace, in which four hydrogen atoms are fused into one helium atom, leading to temperatures in the Sun's outer layers that are roughly two to four million degrees. The surface is about 10,000°F. Over 600 million tons of hydrogen are converted into helium every second. The missing four million tons of mass are converted into pure energy in the form of heat and the motion of atoms, according to Einstein's famous equation, $E = mc^2$.

The Sun, ultimate source of all life and energy here on Earth, is nearly a million miles across and is about 333,000 times the mass of the Earth. If we could stand on the Sun, we would weigh about twenty-eight times what we weigh here on Earth. The Sun is a Class G, or yellow, star and is about halfway through its stable part of its life. Scientists who have analyzed the Sun's light have found about sixty elements present.

Galileo, observing sunspots through his thirty-power telescope in 1610, noticed that the Sun rotates. The Sun is not a solid body like, say, a billiard ball. It is more like a fluid, and all parts or areas of the Sun do not rotate together. The equator takes twenty-five days to rotate once, but the polar regions need more than thirty days to rotate.

The Sun is in a state of equilibrium, remaining the same size by a balance of two forces. Radiation, in which streams of protons are trying to push the gases outward, tries to make the Sun bigger. Opposing this force is the Sun's tremendous gravity, which is trying to make the Sun smaller. Overall, the inward force of gravity is balanced by the outward force of gas and radiation pressure.

The thermonuclear reactions that occur in the Sun can be created here on Earth by a hydrogen bomb, or H-bomb. The first thermonuclear, or "sun," bomb was detonated in 1952. When it was discovered that the temperatures created by a fission atomic bomb were four or five times the temperature at the center of the Sun, we knew we were just a step away from a fusion hydrogen bomb. Scientists are now working on fusion power, but it will probably be at least a few decades before we have a practical and economical fusion reactor.

103. Why, when there is a clear sky and a full moon, do you sometimes see a ring around the Moon?

We've all seen these beautiful rings around the Moon and the Sun. We called them Moon dogs or Sun dogs when we were growing up on the farm. I don't know how the "dog" got in there.

High, thin, cirrus clouds contain millions of tiny ice crystals. The ice crystals are kept aloft by rising air currents. Each ice crystal is an elongated hexagonal, or six-sided, shape and acts as a miniature lens. Light enters one crystal face and exits the opposing face refracting, or bending, twenty-two degrees, which corresponds to the radius of the Moon or Sun halo. Most of the ice crystals are about the same size and shape, so the Moon ring is always about the same size. The ring can even appear to have colors such as in a rainbow.

A friend of mine who's an expert of weather folklore says that a ring around the Moon means bad weather is coming. The high, thin, wispy cirrus clouds that cause the ring normally precede a warm front by one or two days. Normally, a warm front is associated with a low-pressure system, which means stormy weather.

When a low-pressure system is over a region, the air begins to rise. Rising air cools, and the moisture condenses into precipitation and clouds. So a low-pressure system means an increased chance of clouds, rain, and snow. A nice way to remember: L is for low or lousy. A high-pressure system usually brings good weather; H is for high or happy.

Once in a great while, under conditions of rising air currents, the majority of the ice crystals will have a different orientation. Then the ice crystals produce a halo of about forty-six degrees, which makes the rings dimmer than the twenty-two-degree halos. This rare type of ring is seen around the Sun more often than around the Moon.

104. How do satellites always stay in the same place in the sky?

Not all satellites do—only the ones called geosynchronous. From our Earthling point of view, a satellite in geosynchronous orbit appears to hover over one point on Earth. The satellite is in a high orbit when it circles the Earth once a day, the same amount of time it takes the Earth to rotate on its axis. A receiving dish on Earth can point to the satellite at one spot in the sky, and the dish does not have to move to track the satellite. There are now about a couple hundred satellites in geosynchronous orbit located above the equator.

A set of rules, called Kepler's laws, determines how far a satellite must be above the Earth's surface for a geosynchronous satellite. The closer an orbiting object is to the Earth, the faster it goes and the less time it takes to orbit. The space shuttle (when it was in service) and the International Space Station (ISS) are 250 miles high and take about ninety minutes to orbit. A geosynchronous satellite, at 22,200 miles away, takes twenty-four hours, or one day, to orbit. In comparison, the moon is 240,000 miles up and takes about thirty days to orbit.

Early satellite dishes for home use were about eight to twelve feet across. We still see a few of those around. About ten years ago, technicians put up a very powerful satellite transmitter. So the receiving dish could be much smaller, typically about eighteen inches across.

I highly recommend the website heavens-above.com, which shows the exact location of the ISS and Hubble Space Telescope, as well as the positions of the planets. It will show the night sky for your location, discuss meteor showers, and list Iridium flares that you can see, too. Iridium flares are bright flashes of light in the sky seen in the early morning or early evening, caused by reflections of the Sun off the shiny flat antennas of some sixty-six satellites owned by Iridium Communications.

105. What is a black hole?

A black hole is a place in Washington, DC, where all our tax dollars go. Just kidding! A black hole is perhaps the strangest object in the universe. It is the remains of a massive dead star that has run out of fuel and collapsed.

There are two main, competing processes that shape stars. Fusion reactions are similar to tiny hydrogen bombs going off and tend to make the star bigger. At the same time, gravity tends to crunch all the solar material to the center. These two forces are balanced throughout a star's life, which typically lasts for billions of years. The size of a star is determined by this balance between gravity, making it smaller, and explosive forces, making it bigger, which shifts only at the end of a star's life, when the ultimate fate of any star is determined by its mass.

Here's what happens to a star the size of our Sun. When nearly all the hydrogen is converted to helium, gravity will dominate and the Sun will collapse, ignite the nuclear ashes of helium, and fuse them into carbon. The Sun will then expand to the size of the orbit of Mars, at which point it will be a red giant. After a few million years, the helium will be all burned out, the red giant will collapse, and the Sun will become a cool cinder, called a black dwarf. Our Sun will never be a supernova. It is just too small.

The story is quite different for a star more than ten times the mass of the Sun. A massive star can become a supernova. Once nuclear fusion is done, the star's gravity takes over, pulling material inward and compressing the core. This compression generates heat, eventually leading to a supernova explosion, which blasts material and radiation out into space. But it doesn't end there. If the star's mass was around ten times our Sun's mass, then the small, dense core remaining after the supernova keeps collapsing until its protons and electrons fuse together to create neutrons, and an extremely dense neutron star is born. If the star's mass was even greater, the star not only caves in on itself, but the atoms that make up the star collapse so there are no empty spaces. What is left is a core that is highly compressed, very massive, and very dense. Gravitation is so strong near this core that even light can't escape. The particles within the core have collapsed and crushed themselves out of visible existence. The star disappears from view and is now a black hole.

So if we can't see a black hole, how do we know they exist? By their gravitational effects. Though they're not visible, we can detect or hypothesize about their presence by studying surrounding objects. Astronomers can see material swirling around one or being pulled off a nearby visible star. The mass of a black hole can be estimated by observing the motion of nearby visible stars.

The core, or nucleus, of Galaxy NGC 4261, for example, is about the same size as our solar system, but it has a mass 1.2 billion times as much as our Sun's. Such a huge mass for such a small disk indicates the presence of a black hole. The core of this galaxy contains a black hole with huge spiral disks feeding dust and other material into it.

What happens inside a black hole? Much is unknown. When astrophysicists talk about the workings of black holes, they speak a different language: event horizons, singularities, gravity lenses, and ergospheres—strange stuff! And yet, incredibly, Albert Einstein predicted the existence of black holes way back in 1915 in his General Theory of Relativity.

106. What is a shooting star, or falling star?

A shooting star actually comes from our own solar system that has only one star, namely the Sun. A shooting star is a meteor that passes through our atmosphere and becomes extremely hot due to friction. It gives off light.

Meteors enter our atmosphere at speeds of between 25,000 mph and 160,000 mph. Those tremendous speeds cause a lot of friction between the meteor and air. The meteor burns with an extremely high temperature and emits light, much like the filament in a lightbulb. We label that streak of light across the night sky a meteor.

They're called meteoroids when they travel though space. The term "meteor" is used to describe the object when it makes a lighted visible path through the atmosphere. If a piece is big enough to survive these tremendous temperatures and actually hit the Earth, it is termed a "meteorite." You can find meteorites in museums.

Thank God we have an atmosphere. Not only does our atmosphere provide us oxygen for breathing, but it also acts as a buffer zone, protecting us from meteor impact. Very few meteors reach the Earth's surface, with most turning into vapor before they hit Earth. It's a different story on the Moon. The Moon has no air, so it has no atmosphere. And so the Moon's surface is pockmarked with meteor impacts. The Earth does have some notable craters of its own, however. The Tunguska object was a meteor impact in Siberia in 1908 that flattened trees up to fifteen miles away.

Most meteors are observed when the Earth passes through a part of its orbit where a comet passed before. Comets are debris left over from the formation of the solar system. They have large elliptical, or oval, orbits. Comets swing around the Sun in predictable cycles. Probably the most famous is Halley's Comet, which visits us about every seventy-six years.

As comets go around the Sun, they shed an icy, dusty debris stream. The Earth passes through this junk, and bits and pieces of the comet ignite in the searing friction of the atmosphere. They become so hot they give off light, hence the name shooting star. Most are dust- or pea-size. They are traveling at thousands of miles per hour. Moving faster than sound travels, they give off a sonic boom.

Meteor hunters go out looking for them. The first recovery from the April 14, 2010, meteor was made by a dairy farmer near Livingston, Wisconsin. He said he was "drinking a beer in his chair" and the meteorite exploded above his house and a piece hit his shed and bounced right next to him. He found it in his driveway. A meteor collector from Illinois paid him two hundred dollars for it. That's good "beer money." The meteor, witnessed by hundreds of people in the Midwest, could have been a part of the Gamma Virginid meteor shower that began on April 4. It could also have been a rogue one. The April 14, 2010, meteor was believed to be traveling at 36,000 mph.

The best collection of meteorites in the world is at the Field Museum in Chicago. Visit the Barringer Crater in Arizona, where a meteor one hundred and fifty feet in diameter struck fifty thousand years ago. The crater is about three-fourths of a mile across and more than 500 feet deep. Astronauts used it for training for lunar surface exploration in the 1960s. The largest meteorite found in the United States is the fifteen-ton Willamette Meteorite, found in Oregon.

The rocks and dust that fall as meteors come from the Earth's passing through the debris of a comet. Meteor showers are an increase in the number of meteors at a particular time of the year. They are named after the star constellation from which most seem to fall. The best-known and most prolific meteor shower is the so-called Perseids, seen in the direction of the constellation Perseus; this occurs around August 12 of every year. A viewer can see several dozen of these falling stars in an hour.

The first known case of a human hit by a meteorite was Ann Hodges in Alabama. In 1954, an eight-pound stony-iron meteorite crashed through her roof, bounced off her radio, and badly bruised her. According to rumors, she went to church that very day!

107. Why do we bother to send people into space?

George Mallory was an English mountaineer who was part of three British expeditions to climb Mount Everest. He and his partner, Andrew Irvine, disappeared on the northeast ridge in 1924. Someone had asked Mallory, "Why do you want to climb Mount Everest?" He reportedly replied, "Because it's there." Mallory's body was found on May 1, 1999.

That answer may also be a good enough reason to travel into space, orbit the Earth, voyage to the Moon, and venture on to Mars and beyond. We have forever looked to the heavens and wondered what was there. Humans have an insatiable appetite to explore.

However, the human race's first steps into space, in the 1960s, were not motivated primarily by a quest to explore the heavens. Instead, the reasons were mainly political. We, meaning the United States, were locked in a fierce struggle with totalitarian Soviet Communism in what was known as the Cold War. Which system of government was the best, socialism or free enterprise? Who had the best technology? The rest of the world was watching.

The Soviet Union beat the United States into space by launching the first man-made artificial object to orbit the Earth, called *Sputnik*, on October 4, 1957. The United States responded with *Explorer* in January 1958.

The Russians had many firsts: the first human into space (Yuri Gagarin), the first space walk, the first space station, first satellite to orbit the moon, and first pictures taken of the far side of the moon.

But the United States caught up with and surpassed the Soviet Union by landing three men on the moon in July 1969. Soon after, space missions became less competitive. The next "first" was when the United States and Russia cooperated in 1975 by having three American astronauts and two Russian cosmonauts link up during Earth orbit.

We may not have realized it at the time, but many good things would come out of that space race. The technology needed for spaceflight has produced thousands of spin-offs that contribute to our national security, economy, productivity, and lifestyle. Foremost might be the Earth satellites used for weather forecasting and Earth Resource Satellites (ESA) to monitor crops, flooding, pollution, insect infestations, crop yields, and the health of forests.

It is difficult to find any area of everyday life that has not been improved by these technological advancements. Microcomputers, design graphics, compact discs, whale identification practices, the development of earthquake prediction systems, air purification methods, smokestack monitoring, devices to measure radiation leaks, scratch-resistant lenses, flat-pane television, high-density batteries, the GPS system, and noise-abatement techniques all result from NASA research.

So why do we explore space? Perhaps it was Edmund Hillary—the New Zealand mountaineer and explorer who, along with his Sherpa guide, Tenzing Norgay, were the first humans to reach the summit of Mt. Everest, in 1953—who said it best: "It is not the mountain we conquer, but ourselves."

4

Technology

108. How was the Internet invented?

The Internet never began as a single creation or entity, like the invention of the telephone. It was an outgrowth of ARPANet, a project of the Advanced Research Projects Agency (ARPA), of the Department of Defense, an attempt to link our military, defense contractors, and universities together.

Prior to the Internet, computer networks were "hooked up" in a linear fashion, and every network had to be operating. If you had three network computers in a row and the middle one went down for repair, the first and last computer couldn't talk to each other.

In 1969, ARPA set up the first network that was not centralized; there wasn't any single computer running the show. With the new system, any number of computers could talk to each other and information was automatically rerouted should any computer go off-line. The fledgling network linked four universities together, with computers from the National Science Foundation also joining in.

By 1983, they switched to the Transmission Control Protocol/Internet Protocol (TCP/IP), still in use today. TCP/IP specifies how data is formatted, addressed, transmitted, routed, and received at the destination. The word "Internet" appeared at about the same time, and commercial activity on the Internet started shortly after. The first countries to participate outside the United States were England and Norway.

How big is the Internet? Very big! An estimated four out of every five Americans will use the Internet this week. Worldwide, about two billion, or about two of every seven people on planet Earth, will log on. The biggest users of the Internet are China, the United States, India, Japan, and Brazil. Of course, that has much to do with the population of these countries. If we go by the percentage of the population that use the Internet, four countries, Japan, United States, France, and Korea are about the same, right around 80 percent.

The World Wide Web is the most popular part of the Internet. It started in 1989 with fifty people sharing Web pages. Today it seems that everybody and her brother has a website.

Advertising is also big business on the Internet. And at this point, many kinds of companies do *most* of their business on the Internet.

Now that the Internet has grown so large, there are around twenty major search engines out there to help us navigate it, and competition is fierce. The top five, in order of popularity, are Google, Bing, Yahoo!, Ask, and AOL.

109. How do helicopters steer?

A helicopter is an amazing machine. A train can go forward and backward. A car goes forward and back plus left and right. A plane can move forward, left and right, and up and down. A helicopter can do everything an airplane can do and three additional things: A helicopter can hover motionless in the air, rotate in place, and move sideways, backward, and straight up and down.

A helicopter gets its lift in much the same way an airplane does (see page 152). Air that goes over a wing at a faster speed than air below the wing creates less pressure on the top of the wing and more pressure on the bottom of the wing. That greater pressure on the bottom of the wing is called lift. A helicopter has two or more wings mounted on a central shaft, much like the blades on a ceiling fan. Indeed, the propeller on a small aircraft is really a rotating wing. The same Bernoulli's principle applies. Helicopter blades are thinner and narrower than wings on an airplane, because they have to go much faster. Helicopter blades rotate rapidly to create sufficient lift, whereas the wings on an airplane move at the speed of the plane itself.

The helicopter's primary assembly of rotating wings is the main rotor. Increase the angle of those blades, called the angle of attack, and the wings give lift. A gasoline-reciprocating engine—the same type of engine in our cars—or jet engines turn the main rotor.

So far, we have a vehicle that generates enough lift to get off the ground. But as soon as it was airborne, the whole machine would rotate in the opposite direction from the main rotor, in accordance with Newton's law of action and reaction. Rotor blades turn one way (action), the machine turns the other way (reaction).

There are several solutions. One is to put two main rotors on the helicopter, one in the front and one in the back, then turn them in the opposite directions. An example is Boeing's CH-47 helicopters, flown by the military. You see these Chinooks in the news because they've been used in Iraq and Afghanistan.

But the most common solution is to attach another, smaller rotor on the tail and power it with a long shaft, or boom, attached to the engine. The tail rotor produces the force, or thrust, just like an airplane propeller. The rotor is mounted vertically to produce sideways thrust, counteracting the tendency of the vehicle to spin in the opposite direction from the main rotor. By varying the amount of this sideways thrust, the pilot is able to turn the helicopter left and right. The two "rudder pedals" control the pitch, or angle of the blade, of the tail rotor blades.

But it gets even more complex! The main rotor must control the directions up and down and also laterally (sideways). For this, a swash plate assembly does the job, because it allows a pilot to control both movements at once. Pilots use a collective control in their left hands to control the overall pitch of the main blades. This makes the helicopter go up and down. And in their right hands, they hold the cyclic pitch control, which changes the angles of the blades individually as they go around. This makes the copter move in any horizontal direction: forward, backward, left, or right.

Igor Sikorsky, an immigrant from the Ukraine, is credited with the first successful and practical helicopter. First flown in May 1940, his VS-300 paved the way for most modern helicopters, which are still based on his design.

110. Is it possible for a car to run on vegetable oil?

Just about any indirect injection diesel engine can run on vegetable oil—with a conversion kit, of course. An indirect injection engine is one in which the fuel is not injected directly into the cylinders for combustion, as it is in a typical diesel engine. In an indirect injection engine, the fuel is delivered to a pre-chamber where combustion begins.

Vegetable oil falls into the biodiesel fuel category, and with crude oil prices over one hundred dollars a barrel, biodiesel looks more promising. Biodiesel can be made from corn, animal fats, kitchen cooking oil, and soybeans.

Biodiesel is made from biological matter. It is nontoxic and renewable. Gasoline, on the other hand, is made from petroleum. It is both toxic and nonrenewable.

Biodiesel fuels can be used in pure form, but they are usually blended with petroleum-based diesel fuels. B20 is the most common blend, with 20 percent blended to 80 percent standard diesel fuel. B100 refers to pure biodiesel.

Biodiesel has some advantages over gasoline and pure diesel. It helps reduce dependency on foreign oil, is environmentally friendly, lubricates the engine, reduces engine wear, and has fewer emissions. Furthermore, biodiesel is safer. It degrades faster than gasoline, so in a spill, cleanup is faster and easier. Biodiesel has a higher flashpoint than conventional diesel, which means it takes a higher temperature to make it burn, so it's less likely to accidentally combust or explode.

Biodiesel does have some downsides. Even though there is a decrease in particulate matter in emissions, burning it does emit more nitrous oxide (N_2O), which is a greenhouse gas. The new biodiesel fuels can loosen deposits built up in the system, clogging filters and fuel pumps. Some reports show that there can also be a decrease in engine power and fuel economy. Biodiesels can also cost more.

At the present time, it seems that E85 has a toe in the market that biodiesel does not share. E85 is 85 percent ethanol blended with 15 percent regular gasoline. Biodiesel use has been restricted mainly to enthusiasts—people who have, on their own, converted older diesel vehicles to run on cooking oil from deep-fat fryers used in restaurants.

111. How do airplanes stay up in the air?

It seems hard to believe that an aircraft, like the Air Force's C5-A cargo plane, with a gross weight of eight hundred thousand pounds, can stay aloft. How can something so big and so heavy actually get off the ground?

There are two theories of how lift occurs on a wing. Both are correct and both are useful in explaining the forces on an airplane.

The conventional and classic Longer Path explanation uses the Bernoulli effect. The top surface of the wing is more curved than the bottom side. Air traveling over the top of the wing has a greater distance to go—a longer path—than air passing underneath the wing. So the air over the top of the wing must travel faster than the air under the wing. The air passing over the top of the wing and from below the wing must meet behind the wing, otherwise a vacuum would be left in space.

The Bernoulli principle states that faster-moving airflow develops less pressure, while the slower-moving air has more pressure. That

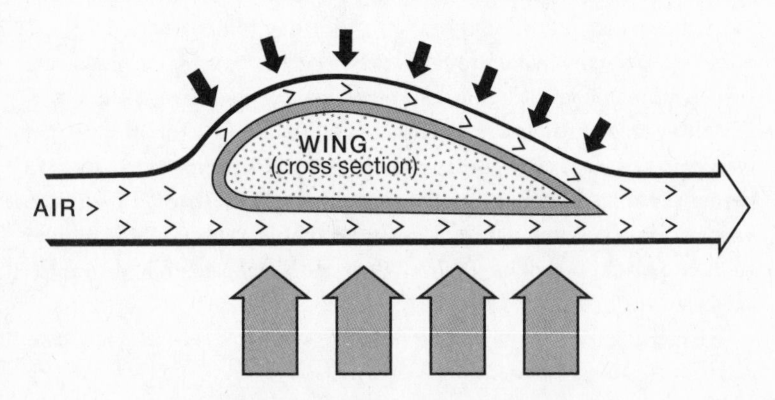

greater pressure on the bottom of the wing is termed "lift." Basically, this pressure difference creates an upward suction on the top of the wing.

The competing but complementary theory is based on Isaac Newton's third law of motion, the law of action and reaction, which states that "for every action there is an equal and opposite reaction." When air molecules hit the bottom surface of the wing at a glancing angle, they bounce off and are pushed downward. The opposite reaction is the wing's being pushed upward. Hence, we have lift. It's similar to BBs hitting a metal plate—they'll rebound backward. BBs go in one direction and the metal plate goes in the opposite direction.

As it turns out, the Bernoulli Longer Path theory is a better explanation of lift for slower-speed planes, including jet airliners. Newton's third law is best suited for hypersonic planes that fly high in the thin air at more than five times the speed of sound.

The navy's Korean War–era Douglas AD Skyraider was the first military plane to carry its own weight. The Douglas DC-7, put into passenger service in the early 1960s, was the first commercial plane to lift a load equal to its empty weight.

112. How can a laser perform eye surgery without hurting the eye?

Eye surgeons have used lasers in retinal surgery for over forty years. They use them to zap bleeding blood vessels on the retina and also to repair a detached retina. The retina is the layer of cells in the back of the eye that receives light and converts it into electrical impulses that are sent to the brain. Eye surgeons also use lasers to correct nearsightedness. The most common case is the LASIK procedure that we see advertised in newspapers and on television. LASIK is an acronym for laser-assisted in-situ keratomileusis.

We call a person nearsighted if their eyeball is too long, so the cornea and lens in their eye focuses light on a point in front of the retina. Nearsighted people see nearby objects clearly, but distant objects are blurry to them. For farsighted people, it's the other way around. Younger people tend to be nearsighted, whereas older people tend to be farsighted.

In additions, older people who were nearsighted when they were young don't necessarily have any improvement in their distance vision when they lose their near vision.

LASIK surgery is very effective for treating nearsightedness. LASIK flattens the cornea, the tough, transparent outer covering of the eye. The cornea does about two-thirds of the refracting, or bending, of light that enters the eye, and the lens does about one-third. Flattening the cornea lets the eye focus light farther back, right on the retina, just like a normal eye does.

Laser eye surgery removes a little of the cornea to flatten out the curvature. In the LASIK procedure, the surgeon folds a flap of the outer corneal tissue out of the way and uses the laser to reshape the underlying corneal tissue, then replaces the flap over the reshaped areas to let it heal to the new shape. Those big, colorful ads in newspapers and magazines do a nice job of showing how the LASIK surgery is performed. Healing is fast, and results are excellent. The cost is several thousand dollars per eye.

What are the problems? There are three main possibilities. One is undercorrecting. The surgeon doesn't remove enough underlying tissue, and the person retains some nearsightedness. The second is overcorrecting. The surgeon removes too much tissue, and the person remains slightly farsighted. The third is wrinkling. The surgeon leaves a small fold or wrinkle under the flap, causing a blurry area in the person's vision.

IBM developed the excimer laser that made laser eye surgery possible. The reason the laser doesn't hurt the eye is twofold. First, it operates in the ultraviolet (UV) region. The wavelengths of the excimer laser beam are just slightly shorter than the waves we can see. Second, the beam can be focused to less than one-hundredth the diameter of a human hair. The beam penetrates less than a billionth of a meter into the surface, which absorbs the UV light on contact, vaporizing the cornea tissue one thin layer at a time.

Select a doctor who has done hundreds of these operations. Success depends on the skill of the surgeon as well as the precision of the instruments. It is highly recommended that you talk to patients who have already had the procedure. Check the reputation of the clinic. Vision is priceless.

113. What is horsepower?

Horsepower is a term that describes how much work an engine or other source does over a specific period of time. The precise definition is that a single unit of horsepower is thirty-three thousand foot-pounds per minute. If you lifted up thirty-three thousand pounds one foot in one minute—that's a rate of 550 pounds per second for sixty seconds—you have been working at the rate of one horsepower.

We give James Watt credit for naming the horsepower. Steam engines were replacing horses in the late 1700s, and there had to be a way to rate the power of steam engines. So Watt calculated the work an average horse could do. He figured a horse could raise 330 pounds up one hundred feet in one minute. That is equivalent to 550 pounds per second or thirty-three thousand foot-pounds per minute.

We measure the power of modern engines by hooking up the engine to a machine called a dynamometer. The dynamometer applies a load to the engine and converts its torque (turning effect) to horsepower. We measure modern car and truck engines in terms of engine displacement, usually in liters. Engine displacement is the total volume inside all the cylinders of an engine. It is difficult to make an exact conversion from liters to horsepower, because there are so many variables, such as number of cylinders, bore diameter, stroke length, and revolutions per minute (rpm) of the engine. My 2005 Dodge Caravan has a 3.3-liter, six-cylinder engine. It is rated at 180 horsepower. So for this car, one liter is about fifty-five horsepower. The popular Ford Focus can come equipped with a two-liter Duratec engine that is rated at 136 horsepower. One liter on the Ford Focus is about sixty-two horsepower. We still use the horsepower rating for many of our smaller engines, such as lawn mowers, snowblowers, leaf blowers, and chainsaws, so people are familiar with the unit.

Humans can generate the equivalent of about one horsepower for a short period of time. Unlike an engine, our muscles tire, so we are limited in the amount of time we can expend power. Each year physics students across the country time themselves running up a flight of stairs. Then they retreat to the classroom and calculate their horsepower.

Here's an example. A 150-pound student runs up a flight of steps that is fifteen feet high in a time of four seconds. That works out to 563 foot-pounds per second, which is a bit over one horsepower. Most students are surprised by how little horsepower they can develop compared to the machines we can build. A good athlete can develop a third of a horsepower for an extended period of time, say, at least a half an hour.

114. Why don't we use more wind power to generate electricity?

We all want sources of electricity that are plentiful, cheap, and nonpolluting. Unfortunately, when it comes to energy production, there is no free lunch. All energy sources have advantages and disadvantages, good points and bad points. There are many factors to consider: availability, effect on the environment, safety, cost, reliability, site selection, and transmission capability.

Using the wind has enticing advantages. Wind power is clean. There are few polluting by-products. The cost of wind power has been coming down so that some wind farms are producing electricity at five cents per kilowatt-hour.

But it certainly is not free, as some would have us believe. If there were potent advantages to wind power, we would be using more of it. So why aren't we? Since the wind does not blow all the time, wind power does not have the reliability associated with coal, nuclear, or natural gas. Remember, we don't have any way to store appreciable amounts of electricity. When it is generated, it must be used immediately.

I've seen acres of windmills in the San Francisco area and also near Palm Beach. It's quite a sight! But if you talk to the local people, they tell you they don't want to live next to them, because they can be ugly and noisy. Wind generators also require a lot of acreage in regions where the wind blows on a steady basis. (Most wind generators need a wind speed of at least 7 mph.) The towers and generators use a lot of steel and aluminum, all of which requires processing and possibly mining, which means more pollution. Also, the generators depend on considerable labor-intensive maintenance.

Wind power has its place. It should be part of the mix. But in the foreseeable future, we will get only a small fraction of our electricity from wind. The total installed wind capacity in the entire United States is sixty thousand megawatts, which is equivalent to sixty power plants.

115. How do remote controls work?

The world's first remote control devices were those used by the German Navy in WWI to steer their radio-controlled motorboats into Allied boats. In World War II, both the United States and Germany used remote controls to detonate bombs. The Era Meter Company of Chicago offered an automatic garage door opener to the public in 1948.

In 1950, Zenith developed a TV remote control appropriately called the Lazy Bones. A long cable attached the remote to the TV set. The remote would activate a motor that would rotate the channel tuner in the set. In those days, tuners were mechanical, not electronic. Eugene Polley invented the first wireless TV remote, called the Flash-Matic, in 1955. Shining a regular flashlight on photocells placed in the four corners of the set operated the functions of on-off, volume, and channel selection. But people forgot which corner to use, plus sunlight would wreak havoc by activating the controls. Then Zenith brought out the Space Command in 1956, which used ultrasonic waves. Unfortunately, clinking metal would affect it and it made dogs bark.

Today, most all remote controls for TVs, VCRs, DVDs, CD players, and home-entertainment systems use light. But it's a kind of light we can't see: infrared (IR) light, whose waves are longer than red, but shorter than radio, television, or microwaves. The IR remote control that we hold is the transmitter, and it sends a binary code that represents the commands we enter. If you look at the end of the remote control that you point to at the TV, stereo, or other such device, you can see the opening, or port, that emits the IR light. There is a receiving device on the TV that picks up the IR light from the transmitter, sorts out the binary signal (a series of 0s and 1s), and carries out the command, such as volume control or channel change. IR remotes are limited to line of sight and have a range of about forty feet, which is sufficient for most purposes.

Remotes to open garage doors use radio waves of between 300 and 400 megahertz (MHz). Key fobs to open car doors and entry doors use 315 MHz for cars made in North America and 434 MHz for cars made in Europe and Asia. The same frequencies are used for most remote-controlled toys. Radio waves can go through materials such as walls and glass, which means the waves can go through our windshield to open our garage door. And the range is usually at least one hundred feet.

Yes, remotes have made life easier and, you might say, "handier." But what about the contention over who gets the remote? Have remotes contributed to obesity because we don't have to get up to change the channels? And terrorists use remotes to detonate bombs and improvised explosive devices (IEDs), another name for roadside bombs.

116. How fast is one Mach in mph?

The Mach number of a moving object is the ratio of the speed of the object to the speed of sound. The Mach number is named after the Austrian physicist and philosopher Ernst Mach (1838–1916).

Sound travels at about 750 mph. So a plane flying at 750 mph is doing Mach 1. If it is moving at 1,500 mph, it is going at Mach 2. Mach 3 would be about 2,250 mph. A speed of 750 mph is about the same as twelve hundred feet per second, or 340 meters per second in the metric system. So it takes about four and a half seconds for sound to travel one mile. This makes it an easy task to calculate how far away from us lightning has struck. Count the time between lightning flash and thunder and divide by five to get the distance in miles. We say "about" because the speed of sound depends on the temperature of the air. Sound travels a bit faster in warm air. So if you want to break the sound barrier, do it on a cold day. You don't have to go quite as fast!

Anything traveling below the speed of sound is said to be subsonic. Supersonic refers to objects moving faster than the speed of sound. The British-French supersonic transport (SST) plane *Concorde*, which is no longer in passenger service, flew at Mach 2. Most of our top-line military fighter planes, such as the F-14 Tomcat (retired from service), F-15 Eagle, F-16 Falcon, F-18 Hornet, and F-22 Raptor, all fly at about Mach 2. The F-117 Nighthawk stealth fighter is listed as 0.92 Mach, just below the speed of sound. The former space shuttle orbited the Earth at a speed of over 17,000 mph, and the current International Space Station still does; this calculates out to Mach 23.

We usually speak of the speed of an aircraft when we use the term "Mach number," but a Mach number can describe the speed of any object traveling through air. When we hear the sharp crack of a bullwhip, it means the tip of the whip is breaking the sound barrier. When someone snaps a towel, a playful but painful locker-room trick, the end of the towel is moving faster than sound moves.

When an aircraft exceeds Mach 1, high pressure builds up in front of the plane. The shock wave spreads backward and outward in a cone shape. It is this shock wave that we hear as a sonic boom.

Before 1947, most people thought that no plane could fly through this so-called sound barrier. Combat fighter pilots in World War II would go into a dive, and the air moving over the top of the wings would exceed the speed of sound; then their controls would freeze up and the plane would crash. Captain Chuck Yeager, at the controls of the Bell X-1, broke the sound barrier on October 14, 1947. Shaped like a .50-caliber bullet, the four-chambered rocket plane was named *Glamorous Glennis*, in tribute to his wife. It now hangs in the Smithsonian Air and Space Museum in Washington, DC.

117. How does Bluetooth work?

Bluetooth is a technology that allows devices such as cell phones, cordless phones, television sets, personal computers, stereos, DVD players, entertainment centers, and TV and satellite radio to talk to each other. These devices can communicate with each other without stringing messy wires and cords between them. It is all done with radio signals.

Bluetooth creates a small area network that operates on a frequency of about 2.45 gigahertz (GHz). The US government has set this frequency band aside for industrial, scientific, and medical use. It is the same band used for baby monitors and some garage door openers and cordless phones.

Bluetooth devices send out very weak radio signals—in some cases, as little as about one milliwatt. The coverage distance is typically about thirty feet. That extremely low power ensures that a Bluetooth device won't interfere with other devices and is also very easy on batteries.

Bluetooth uses a technique called frequency hopping spread spectrum. It uses seventy-nine randomly chosen frequencies, changing from one to another on a regular basis. The transmitters change frequencies sixteen hundred times every second. So it is unlikely that two transmitters will be on the same frequency at the same time.

Bluetooth allows an operator to set up a personal area network (PAN) in their house and car. This "piconet" may tie together their cell phone, stereo, computer, DVD player, satellite TV and radio receivers, and cell phone.

Ease of setup is one of the big attractions of Bluetooth. No need to wade through a ton of instruction manuals. When any Bluetooth-ready device is turned on, it automatically sends out radio signals to other devices. Then it "listens" for radio signals in response. Once it identifies a signal, it locks it in and remains active with other devices within that thirty-foot distance.

Bluetooth allows hands-free cell phone use. That's a big advantage for car drivers, as many states have passed laws banning cell phone use and/or texting while driving. Users can be jogging or fishing or relaxing on the veranda while listening to music on their Bluetooth. The jawbone-shaped headset is now a common sight.

Bluetooth was developed by Ericsson, a large telecommunications company in Sweden. In Nordic lore, Harald "Bluetooth" Gormsson was king of Denmark from 958 AD to 985 or 986 AD, when, according to some, he was killed in a rebellion led by his son. He had united Denmark and parts of present-day Sweden and Norway into a single kingdom, and he introduced Christianity into his kingdom. He erected a large runestone, a stone with an inscription, in memory of his parents in Jelling, Denmark. Thus, the choosing of the name Bluetooth is a testament to the Nordic national pride in having discovered this form of wireless communication and because Bluetooth unites different communication protocols, as King Harald united his kingdom.

118. Why have humans technologically progressed while all other animals have not?

Isn't human progress truly amazing? Seventeenth-century philosopher Thomas Hobbes described the life of early humans as "solitary, poore, nasty, brutish, and short." We humans have come so far in so little time. It is hard to imagine what our daily life would be like if we were back in those times.

The caveman is a popular character based on ideas of how early humans may have looked and behaved. They're usually pictured as hairy creatures clothed in animal skins, armed with clubs and spears, and oftentimes dumb or aggressive. Pop culture promotes this view, for example, in comics like *B.C.*, *Alley Oop*, *The Far Side*, and the animated television series *The Flintstones*. Some of these depictions show cave people living at the same time as dinosaurs. There is strong evidence that dinosaurs died out about 65 million years ago. The only mammals at that time were small, furry, four-legged creatures.

We can picture early humans searching for food, trying to stay warm, protecting their territory, tending to their sick and injured, learning to use tools, and burying their dead. It must have been a nearly full-time struggle just to stay alive. Two developments allowed our early ancestors to dominate their environment: growth of the frontal lobes of the brain and evolution of the opposable thumb. These were instrumental in the human race's long progress from cave to castle. The frontal lobes gave us the capabilities of impulse control, judgment, language, memory, motor function, socialization, and problem solving. They are responsible for planning, controlling, and executing behavior. In other words, the frontal lobe gives us all the functions we ascribe to humans. I recommend the Carl Sagan book *Broca's Brain*, which describes how the frontal lobe is concerned with reasoning, planning, parts of speech, movement (motor cortex), emotions, and problem solving. Another fine read is Steven Jay Gould's *The Mismeasure of Man*.

The opposable thumb allowed us to make and manipulate tools with great dexterity. If you are tempted to disregard the importance of an opposable thumb, try these tasks: tie your shoelaces, or blow up a balloon and tie it, without using your thumbs. Stereoscopic vision gave the tool of depth perception, a great aid in using tools and in hunting.

Another key feature of our long march of progress is the role of a written language. Knowledge exploded when people were able to permanently record what they learned and pass it on to the next generation. The oral tradition is very limited, and could be somewhat responsible for the differences in the development of civilization between Europe and North America in the time leading up to the 1500s.

But there is an element of the human ascent that is troubling: our penchant for destruction of our own kind and of our environment. Warfare seems to be the scourge of humanity. Many of us now have unprecedented wealth, comfort, longevity, medical care, and leisure time. Yet we have trouble getting along with each other. To make things worse, we have the capability to destroy other humans on a massive scale. It is "only the dead who have seen the end of war." Some attribute this quote to Plato, an ancient Greek philosopher (428 BC–348 BC), while others say it's from the writings of Spanish-American writer George Santayana (1863–1952). Whatever the case may be, I'm afraid the phrase may hold some truth.

What is the future of science and humankind? Science is a cultural pursuit, in the same vein as music, art, and literature. It is difficult to comprehend the mysterious beginnings of atoms, stars, galaxies, and planets. It is humbling to be able to understand how life emerged, advanced, and developed into a biosphere containing creatures like us, that have brains able to ponder the wonder of it all. This common understanding should rise above all national differences and all religions. Human beings are the only creatures that can ask the questions "Why am I here?" and "What is the meaning of my existence?" and "What does the future hold?"

But that complexity comes at a price: ignorance of the things around us. Many modern gadgets are like magic black boxes. Take apart a cell phone, look at the miniaturized parts, and there is not a clue as to how it works. That intricacy and complexity will only increase. Some people discuss implanting chips and circuitry in the brain that will augment its computing power, making us part human and part machine.

There is every reason to believe that the steady progression of scientific knowledge and implementation of that information will continue. It is progress that gives ever more people access to greater comfort, food, clothing, and shelter and to longer life, increased leisure time, and easing of pain. But we must be aware of the limitations of science. Just because we know more, that does not mean we behave any better or treat others in a humane manner. Some people knowingly ingest carcinogens, take illicit drugs, and drink too much alcohol. Some people are in poverty because they make bad choices, or because of others' self-interested choices. We *know* so much better than we *do*.

We are great at solving technical problems. Putting a man on the moon was an engineering task. We understood rocket propulsion, celestial mechanics, navigation, and life-support systems. Our country had the money, will, and focus to succeed. But we are not very good at "people" problems. We have difficulty convincing people to make good decisions in use of food, alcohol, driving, drugs, etc. A free society cannot compel people to live or behave in a healthful manner. It must persuade them that it is in their best interest. And that is not an easy task. Just because science and technology can give us all these "goodies" does not mean people use them in a constructive manner. They don't necessarily make people happier or lead them to live more productive lives. Scientific progress does not advance equally in all areas of human endeavor. The war on cancer is improving. But there remains much that we do not understand about cell biology. We've been able to predict lunar and solar eclipses centuries in advance at least since the time of the ancient Mayans. But predicting clear skies or cloudy skies only one day ahead is iffy. There is room for pessimism and for optimism. On the dark side, there are groups that will use science and technology to commit mass murder and eliminate those who hold ideas different from their own. On the positive side, there is hope that third-world countries will narrow the gap with the developed countries and enjoy the prospects of clean air and water, sufficient foodstuffs, and the freedom of choice we have in the Western world.

119. Why do magnets pull together?

In 1883 or 1884 (accounts differ), Hermann Einstein brought home a compass and gave it to his four- or five-year-old son, Albert. Young Albert Einstein observed that the compass needle always pointed in the same direction and moved without anything touching it. He later stated it was at that moment he realized "something deeply hidden had to lie behind things."

Like most compasses, Albert Einstein's compass was just a bar magnet allowed to swivel on a pivot. Magnets truly are fascinating devices. We know much about how magnets operate, yet much about magnetism, especially at the atomic level, remains a mystery more than one hundred years after Albert Einstein played with that simple compass.

We do know that a magnet is any object that has a magnetic field around it and attracts objects made from iron, nickel, and cobalt and their compounds; these substances exhibit the property of ferromagnetism. ("Ferro-" comes from "*ferrum*," the Latin word for iron.) The Greek philosopher Thales of Meletus recorded that natural magnets were found in the region of Magnesia, hence the name magnet.

An early word for magnet is "lodestone," a Middle English word meaning "leading stone" or "course stone," from its use in keeping ships on course. Ship navigation was often accomplished by placing a natural magnet (lodestone) on a piece of wood and then floating the combination in a bucket of water. The lodestone acted as a crude compass.

The most common magnet is a bar magnet, often marked "north" on one end and "south" on the other. Another familiar magnet is the horseshoe magnet, which is merely a bar magnet bent around into the shape of a horseshoe. A horseshoe magnet is stronger than a bar magnet the same size, because both poles can pull on metal objects instead of just one pole.

"Like" poles, such as north and north or south and south, repel each other. "Unlike" poles, north and south, attract each other. Magnets are strongest at their poles. The force of attraction varies according to the inverse square law: twice the distance away, one-fourth the pull. The north pole of a free-swinging magnet points to the geomagnetic North Pole (which is magnetically a south pole, even though we call its direction magnetic north), which is located up in Hudson Bay above the Arctic Circle. The position of the magnetic North Pole moves around from year to year.

But why are iron, nickel, and cobalt the only elements attracted to a magnet? These atoms have unpaired electrons in their outer orbits. Not only do these electrons go around the nucleus, they also spin on their axes, something like the way the Earth spins on its axis. The spin on an electron creates its own tiny magnetic field. If all the electrons are spinning with their axes pointed in the same direction, each one will exert a little tug, or pull. Adding them all together, you get a big tug. Groups of these atoms in which the electrons are spinning in the same direction are called domains. In the presence of a strong magnetic field, created by a current, the effect moves up a level: more and more of these domains align in the same direction. After the strong magnetic field is removed, the magnetic domains of materials that are not good magnet candidates revert to a random orientation. The domains of strong, or good, magnetic materials remain in a structured orientation; thus a permanent magnet is born.

Loudspeakers typically use an alnico magnet, a combination of aluminum, nickel, and cobalt. Refrigerator magnets are usually ceramic. Neodymium magnets are some of the strongest permanent magnets and are used in high-quality products such as microphones, professional loudspeakers, in-ear headphones, and computer hard disks, where low mass, small volume, or strong magnetic fields are required.

Hans Christian Oersted, a Danish physics teacher, discovered the relationship between electricity and magnetism. On April 21, 1920, Oersted was giving demonstrations to students in a classroom. He laid a wire carrying current near a compass and noticed that the compass needle moved. The needle deflected when electric current from a battery was switched on and off. Oersted had been looking for a relationship between electricity and magnetism for several years. The operation of all motors and generators is based on the association between magnetism and electricity.

120. How do Legos connect together?

Legos have reached cult status in the toy world. The popularity of the colorful interlocking plastic bricks has surpassed that of Erector Sets, Lincoln Logs, and Tinkertoys.

Ole Kirk Christiansen, a Billund, Denmark, carpenter, started making small wooden toys and playthings in 1934. Ole, a stickler on quality, named his company Lego, after the Danish expression *leg godt*, meaning "play well." In Latin, "*lego*" means "I put together."

In 1949, Legos came out with and early version the familiar plastic stackable hollow rectangular blocks with round studs on top. The precision-made blocks would snap together and hold, but not too tightly. A three-year-old could pull them apart.

Like many plastic devices, Legos are made by injection molding. The process starts with heating the acrylonitrile butadiene styrene (ABS) plastic the bricks are made of to 450°F to give it the consistency of hot fudge, then forcing it into a mold at very high pressure and allowing it to cool for fifteen seconds. A mold is a steel or aluminum cavity having the shape of the desired part. The finished figure is ejected out of the mold.

Lego pieces have studs on top and tubes on the inside. The brick's studs are slightly bigger than the space between the tubes and the walls. When bricks are pressed together, the studs push the walls out and the tubes in. The material is resilient and wants to hold its original shape, so the walls and tubes press back against the studs. Friction prevents the two bricks from sliding apart.

The popularity of the Lego system stems from its versatility. The design encourages creativity, as children can construct any number of machines, figures, and devices, then tear them down and reuse the parts.

In the late 1960s, Lego introduced the Duplo line of products. These bricks are twice the length, width, and height of the standard Lego block. Often dubbed Lego Preschool, the size of the larger blocks discourages tykes from swallowing them, and they are easier for small hands to manipulate. The Lego people have introduced numerous ancillary enterprises, including six amusement parks, competitions, games, and movies. They also put out new lines of Lego sets, Clikits and Belville, which are targeted to young girls. The addition of gears, motors, lights, sensors, switches, battery packs and cameras are incorporated in the Power Functions Line. Those smart Danish people have made sure that newer products are compatible with the older brick-type connections.

Lego claims they have manufactured 400 billion blocks in the last fifty years. That amounts to about fifty-seven Lego blocks for every person on planet Earth.

121. How do rockets work?

Rockets can be simple or very complex, depending on their size. On a small scale, they're so simple that you and I can order parts from a hobby store, build a rocket, and send it up into the wild blue yonder ourselves. On a more massive scale, rockets are so complex that only a handful of countries, with billions of dollars in resources, have been able to use them to send humans or other objects into space.

Rockets work on the principle of Newton's third law: For every action, there is an equal and opposite reaction. We can experience this fundamental law by letting air out of an inflated balloon. The action of the air rushing out of the balloon propels the balloon in the opposite direction. Rockets work the same way. The gases produced when they burn either liquid or solid fuel rush out the back—that's the action. The reaction is the rocket's moving in the opposite direction. The thrust depends on several factors, including the type and amount of fuel and the length of time it's burned.

The concept of momentum also comes into play in rocketry. Momentum is defined as mass multiplied by velocity. The mass of the gases multiplied by the velocity at which they are ejected is exactly equal to the mass of the rocket times its velocity; therefore, by definition, the momentum of the ejected gases is the same as the momentum of the rocket going up. Even though the gases have very little mass but tremendous velocity, while the rocket has a huge mass but a small velocity compared to that of the gases, the product when you multiply them is the same.

Model rockets are simple and safe. They utilize solid propellants that burn quickly without exploding. You can store model rocket engines for many years and they won't deteriorate, and the fuel does not have to be handled. On the opposite end of the spectrum is the most powerful rocket ever built, the *Saturn V*, which sent Americans to the Moon in the late 1960s and early 1970s. Today, the most powerful rocket is NASA's *Delta IV*, which can send fourteen tons into orbit around the Earth. The first use of the *Delta IV*, which burns liquid hydrogen and liquid oxygen, was in June 2012, to put a spy satellite into space.

The Chinese are credited with developing the very first rockets in the early 1100s; they were an extension of their development of gunpowder.

It's widely believed that the Chinese used rockets against the invading Mongol hordes in 1232 AD. In the early 1800s, William Congreve, in conjunction with England's Royal Arsenal, prepared an efficient propellant mixture and made a rocket motor with a strong iron tube and a cone-shaped nose. During the War of 1812, Congreve rockets, launched by the British Navy, rained down on Fort McHenry near Baltimore in September 1814. This was the origin of the "rocket's red glare" that Francis Scott Key alluded to in his poem "Defence of Fort McHenry." The words were later used in our national anthem, "The Star-Spangled Banner."

Ion propulsion is something quite new and very exciting. In ion engines, electrons from xenon gas are stripped off the xenon atom with a given electrical charge. Magnetic fields fire the ions out the back of a thruster. Ion engines do not use heavy, bulky fuels, and while they generate very low thrust, they can be "burned" for many hours. This makes them valuable for deep space travel and for sending payloads from a low Earth orbit to a geosynchronous orbit, which is the distance at which satellites like those used by Dish Network for TV transmission are in orbit (see page 140). NASA is working on ion propulsion systems with higher speed and more thrust.

122. How is Styrofoam made?

Styrofoam is actually a Dow Chemical trademark for a foamed plastic made from polystyrene, derived from crude oil, which Dow introduced in 1954. Early Styrofoam was formed by blowing hazardous chlorofluorocarbons (CFCs) into a glob of resinous material made of styrene.

Styrene is an organic compound made from benzene, a colorless oily liquid that evaporates easily. Styrofoam is a trademark name for expanded polystyrene that goes into disposable coffee cups, coolers, and cushioning material in packaging.

But most of the world phased out CFCs in the later 1980s because of concern about these gases destroying the ozone layer that protects us from excessive exposure to ultraviolet radiation. Now manufacturers make Styrofoam products by blowing pentane or carbon dioxide gas into a plastic melt. Heating it causes the gas to expand in bubbles of the polystyrene, forming the foam the product is named for. As the resin cools it traps these tiny bubbles of gas inside, forming a cellular plastic structure. Polystyrene foam products are about 98 percent air. That's why they're so lightweight and have such excellent insulating properties.

Some of the most common Styrofoam products are coffee cups, egg cartons, meat and produce trays, soup and salad bowls, "peanuts" used in packaging, and lightweight molded pieces that cushion appliances and electronics in their packaging for shipping.

Tests prove that the use of disposable Styrofoam plates and cups in schools and hospitals prevents diseases. Styrofoam resists moisture and remains strong and sturdy over long periods of time. Styrofoam can be molded into shapes easily, and it costs less than paper products.

Styrofoam has taken some knocks over the years. It can fill up landfills quickly. The polystyrene industry will tell you that Styrofoam makes up less than one percent of the weight of solid waste in a landfill. What they don't say is that the percent by volume is much higher. Other disadvantages of Styrofoam are that it is not biodegradable or easy to recycle, and that burning the stuff gives off toxic gases.

Like so many products in our modern life, Styrofoam has its good side and its bad side. It seems to me that the positives outweigh the negatives.

123. What types of fuel besides gasoline can be used to run cars?

There are five main alternatives to gasoline, each with its own set of pluses and minuses. But the big advantage of all five is that they are less polluting and produce lower amounts of greenhouse gases than gasoline or diesel fuel.

The one we hear the most about is ethanol, or ethyl alcohol. E10 fuel is 10 percent ethyl alcohol and 90 percent gasoline. Most any car sold in the United States can use E10. The cost of E10 is about ten cents higher than regular gasoline in most areas. E85 is 85 percent alcohol and 15 percent gasoline. Not just any old car can use E85. It must have an engine specifically designed for E85; cars that use these engines are called flexible-fuel, or flex-fuel, vehicles. This kind of car can run off regular gasoline if E85 is not available.

E85 can be produced locally, causes less engine knock, and is usually considered less polluting as gasoline. In some places it costs less than regular gasoline, while in others it costs more. E85 is not widely available and reduces fuel economy compared to regular gasoline. About 3 percent of US cars are E85 powered.

Biodiesel is made from vegetable oils or animal fats (see page 151). It can even be made from used cooking oil, which can sometimes make it smell like french fries. B5 is 5 percent biodiesel, and B100 is 100 percent biodiesel. The most popular blend is B20, or 20 percent biodiesel, promoted by singer Willie Nelson. He calls it BioWillie. Biodiesel is not widely available and tends to solidify into a gel in bitter-cold winters, and car manufacturers won't warrant their cars for anything higher than B5. Despite these drawbacks, experts say there may be a place for biodiesel fuels in the future.

Natural gas, in the form of compressed natural gas (CNG) or liquefied natural gas (LNG) is the third alternative to gasoline; it became an option with Honda's Civic CNG, which began production in 2008. Natural gas is far less polluting to burn than gasoline, so it's attractive for use in taxicabs and utility vehicles in crowded cities. Natural gas can be cheaper than gasoline, but the gas tank has to be rather large, taking up the whole trunk space. The rate of miles driven per fill-up is dismal, so you won't see natural-gas cars in wide-open country.

Propane is the fourth possibility. It burns quite cleanly in engines and is also less polluting than gasoline. Propane, in the form of liquefied petroleum gas (LPG), is ideal for fleets of taxis, buses, delivery vehicles, and forklifts. Zion National Park in Utah has a fleet of thirty buses powered by propane. LPG is stored as a liquid under about two hundred pounds per square inch of pressure. Canisters of LPG are also popular for barbecue units and camp stoves. Liquid propane flashes to a vapor when no longer under pressure. A quick release, say, from a ruptured tank, yields a fine white mist as moisture in the air condenses on the gas particles.

It should be pointed out that LPG is usually considered to be 100 percent propane in the United States and Canada. However, in Europe, the propane content of LPG can be less than 50 percent. Propane is the third on the list of most widely used motor fuels in the world.

The fifth alternative fuel is hydrogen, which has enormous potential, mainly because there is no pollution to speak of, since the by-product of combustion is water vapor. Hydrogen can come from water and is used in fuel cells for electric cars, but it can also be used to power internal combustion vehicles. Hydrogen has the potential to be produced locally and anywhere in the world, but the technology is simply not quite there yet.

To be sure, there are significant drawbacks. Fuel cells are expensive. Storing hydrogen in a car is also dicey, the cost is high, and there is no widespread distribution system in place.

124. Why do train tracks buckle during hot weather and cause derailments?

You see it on the news every summer. Someplace on this good planet Earth, train tracks buckle during a spell of hot weather, causing a train wreck. The reason is well understood. Metals expand when heated; thus, the metal rails in the train tracks get longer. In the old days, there were gaps between the thirty-nine-foot rails that made the familiar clickity-clack sound as the train went over the tracks. A thirty-nine-foot train rail can expand as much as a half inch when the temperature goes from 70 to 100°F. The gaps allow the rails to expand.

This process changed a few years ago. Railroad companies now have tracks laid in segments of about fifteen hundred feet or more. This so-called continuous track, or continuous welded rail, eliminates the expansion joints. At the mill, shorter tracks are welded together and carried lying on two flatcars to the intended site. There, the sections are welded together in a thermite fusion-welding process, which involves placing a mixture of aluminum powder and iron oxide in a crucible and igniting them with magnesium. The combustion raises the temperature of the reaction of aluminum powder and iron oxide to more than 5,000°F, and this mixture is poured into the gap between the two preheated rails to weld them together.

There are advantages to employing rails that may be a half mile in length instead of the standard thirty-nine feet. The longer joined rails provide a smoother ride, less maintenance, and less friction than the shorter rails. Trains can travel at a higher speed.

Railroad companies always schedule the process of laying the long sections of track for one of the locally hottest days of the year, so that the rails are not likely to expand further. On cooler days, the rails will be under tension because they will try to contract and pull apart. But the maximum stress on the rail, even if the temperature drops 100°F, is less than one-fourth the allowable stress for the steel in a continuous track.

As a matter of curiosity, the distance between the rails is 4 feet, 8.5 inches, believed to be the same distance as the wheels of Roman chariots. Ben Hur was actually "riding the rails."

125. How do they decide how long airplane runways should be?

The length of airport runways depends on several factors, including the types of aircraft the airport will serve, the airplanes' itineraries, the altitude of the airport, the speed and direction of the winds, the surrounding terrain, and the proximity of tall buildings and towers.

Large, heavy aircraft need a longer runway to achieve the high speed required to give the wings enough lift for takeoff. As for itineraries, more fuel is needed for longer flights, so a higher takeoff speed is used to lift the greater weight of the additional fuel. Some fully loaded Boeing 747 planes weigh close to a million pounds.

Another factor is the elevation difference between airports near sea level, like the one in San Francisco, and those in the mountains, like Denver's airport. The air in Denver, the Mile High City, is quite a bit less dense than the air in San Francisco, so it provides less lift during takeoff. In addition, the less-dense Denver air provides less air for jet engines, which means the engine is not generating as much power as it would at a lower-level airport.

So an aircraft in Denver must reach a higher speed on the ground to be able to take off, requiring a longer runway to give the aircraft time to reach that speed. As a rule of thumb, the runway length is increased by 7 percent for each one thousand feet of elevation above sea level.

For Denver, which is over five thousand feet above sea level, the figuring goes like this: 1.07 (the runway length needed at sea level plus 7 percent) to the fifth power (adding 7 percent to each new total five times, once for each thousand feet) equals 1.4. That is, we can multiply 1.07 times itself five times and we come up with 1.4. So the runway lengths at Denver should be 40 percent greater than the airport runway lengths at San Francisco. That holds with reality: The new runway at Denver is 16,000 feet long, while the longest runway at San Francisco is 11,870 feet in length. The required takeoff distance for the fully loaded Boeing 747-400 at sea level is 11,100 feet, and at Denver's high altitude, a plane needs nearly 5,000 feet more runway to generate the required lift.

Additional reasons to need a longer runway are based on air temperature and humidity. Warmer air is less dense than cold air, so warm air gives less lift. Pilots refer to this as density altitude. Dry air is slightly more dense than moist air. Ideally, pilots want to take off in air that is low (low altitude), cold, and dry.

126. What were the first guns like?

The first gun was essentially a hand cannon. It consisted of a strong metal tube open at one end and with a hole drilled into the tube near the closed end. The user loaded gunpowder into the tube and rammed a ball, or bullet, down the barrel, then stuck a fuse (or poured a bit of powder) into the drilled hole. Igniting the fuse turned the gunpowder into a gas and pushed the ball out of the open end of the barrel with great speed. Mongolians used hand cannons in the 1200s. The barrel was long, the accuracy was terrible, and those early guns were extremely heavy. Today, these hand cannons are called pistols.

The blunderbuss, a gun associated with the Pilgrims, had a short barrel and a flared muzzle. The wide barrel speeded loading and spread the shot. The ignition mechanism for a gun is called a lock. The lock in a blunderbuss held a slow-burning rope that was ignited ahead of time and then moved into position by the trigger to ignite the gunpowder housed in a pan. This gun had some disadvantages; for example, rain would put out the burning rope. Also, the burning end of the rope glowed at night, betraying the shooter's position.

The matchlock was the first weapon that allowed the shooter to keep both hands on the gun. Just as important, it permitted the shooter to keep an eye on the target while firing the weapon.

The flintlock was a big breakthrough and lasted for over three hundred years. Flint is a kind of rock, and when it strikes steel, it creates hot sparks. In a flintlock, that spark ignited gunpowder in the lock. The flintlock was the soldier's main firearm during America's Revolutionary War.

The percussion cap was perfected by the time of our Civil War. The cap, about the size of a pencil eraser, was a short, hollow tube made of metal and closed at one end, loaded with a chemical compound called mercuric fulminate, which is highly explosive. The open end of the cap fit over a "nipple" from which a tube extended to the powder, waiting to be lit. Pulling the trigger made a hammer strike the cap, setting off the explosive.

The cartridge came along about the time of the Civil War, too. A cartridge has the powder and bullet enclosed in a metal casing, with the powder sitting right behind the bullet. A sharp blow to either the rim or the center of the shell ignites the powder.

Not so many years ago, we marked the two hundredth anniversary of the beginning of the Lewis and Clark Expedition to explore the lands acquired by the Louisiana Purchase. The thirty-three frontiersmen ran out of trading goods, whiskey, and food, but during the twenty-eight-month journey to the Pacific and back, the Corps of Discovery never ran out of weapons, bullets, and gunpowder.

127. What are halogen lights?

A regular incandescent lightbulb consists of a thin, frosted, glass enclosure with a tungsten filament inside. When the bulb is turned on, the filament becomes white hot at about 4,500°F. These lightbulbs are not very efficient, because the process of incandescence emits large amounts of heat. Most of the electrical energy put into the incandescent lightbulb goes into heat and not light. And the bulbs only last about a thousand hours. As the tungsten filament evaporates away, depositing particles on the inside of the glass, the bulb darkens (see page 185).

A halogen lightbulb has the same kind of tungsten filament but is housed in a much smaller quartz glass envelope. Inside the envelope is a gas from the halogen group, such as fluorine, chlorine, bromine, iodine, or astatine. The gas combines with the tungsten that is evaporating from the filament; this causes the tungsten to be redeposited on the filament via a halogen cycle chemical reaction.

How does this halogen cycle work? At lower, moderate temperatures, the evaporating tungsten reacts with the halogen to produce a halide gas. At the higher temperatures close to the filament, the halide gas dissociates, releasing the tungsten and the halogen. The tungsten is deposited back on the filament to be reused instead of accumulating on the glass, and the halogen gas is free to circulate again. This halogen cycle gives the bulb a longer life.

This setup also allows the filament to run hotter, which makes the bulb more efficient. The hotter the bulb, the more light produced per watt of electricity consumed. In a smaller bulb, the filament is closer to the surface, which makes the glass bulb hotter. Regular glass would melt at such high temperatures, so manufacturers use quartz.

It is interesting to know the story of the development of halogen lights. In the 1950s, engineers wanted small, powerful lights to fit on the wing tips of jet planes. So General Electric came up with the halogen light to fill that need.

128. **Why can't we create a perpetual motion device?**

Perpetual motion machines violate the most sacred law in all of science, the law of conservation of energy. A perpetual motion machine would have to have more output than input. In other words, it would have to produce more energy than it uses. You don't get something for nothing!

Every machine has a supply of energy, of which it uses some proportion to do work; it gives off the rest as waste heat. The starting energy equals the sum of the energy used to do the work and the energy expended as waste heat. In other words, energy is conserved. Examples of energy would include heat, mechanical, solar, electrical, magnetic, chemical, and thermal.

The law of conservation of energy states that one cannot create energy from nothing. No one has ever found an exception to this rule. Energy is always conserved—it just changes from one form to another. Those forms include mechanical, electrical, magnetic, thermal, chemical, and nuclear energy.

Any perpetual motion machine would have to take energy at some point, use it for work, and yet return everything to its original state, unchanged, at the end of the cycle. Real-world machines and processes leave things changed permanently. Engineers measure this change as entropy. And the second law of thermodynamics requires that any real process increase the entropy of the universe. For these reasons, no genuine perpetual motion machine currently exists, has ever existed, or could ever exist. As early as 1775, the Royal Academy of Sciences in Paris issued a statement that it would "no longer accept or deal with proposals concerning perpetual motion."

In the past, an inventor could apply for and be issued a patent for a drawing, idea, or scheme. They didn't have to produce or build any real device. Some charlatans would find people to invest in a supposed perpetual motion machine, take their money, and invest it in banks or the stock market. Indeed, the investors got their money back when the perpetual motion machine was shown to be a fraud, but the crooks got the interest on that money. The US Patent and Trademark Office (PTO) issued a decree in 2001 that it will not issue a patent for a perpetual motion machine unless the person applying for the patent demonstrates a working model. One can go on the Internet and find all kinds of so-called perpetual motion machines, but they don't have patents, because none of them actually work.

Consider a bow and arrow. By pulling the arrow back, a person does work in bending the bow. The bent bow has potential energy. When the person releases the arrow, it has kinetic energy in its motion. The arrow embeds itself into a straw bull's-eye target. No more potential energy. No more kinetic energy. But the arrowhead and the straw target have a slightly higher temperature: mechanical energy (in the form of kinetic energy) to heat energy. Nothing lost and nothing gained, only a transfer of energy.

Here's another example: A car uses more fuel, gas, diesel, or alternative, when the air conditioner is running, when the headlights are on, and when the radio is playing. The energy to run these functions comes from the battery, which must get it from the alternator, which is turned by the engine, which burns the gasoline. You can't run the air-conditioning, the headlights, and the radio for nothing.

James Prescott Joule, son of a British brewmaster, was instrumental in developing the law of conservation of energy. He proved that for any amount of work done, a definite quantity of heat always appears. Putting it in scientific terms, the total energy of an isolated system remains constant regardless of changes within the system. That's why there can never be a perpetual motion machine. Even in the modern age, there are people who claim to have come up with these machines that have a greater output than input. Can't happen!

The best known of these perpetual-motion people is inventor Joseph Newman. For years he claimed his "energy machine" produces more energy than it consumed. He appeared on the Johnny Carson show. He took millions of bucks from investors. He applied for a patent, but the PTO rejected his application. He sued them. The National Bureau of Standards was asked to test his claims. Its report came back in June 1986, stating that at no time did the output exceed the input. No free energy or perpetual motion. Go online and view his website, www.josephnewman.com, which is very impressive, and the Web page from what is now the National Institute of Standards and Technology that debunks his claims.

129. How do stitches dissolve?

Dissolvable stitches, also called absorbable sutures, can be used for both internal and external wounds and surgical incisions. There is no need for a follow-up trip to the hospital or doctor to have the stitches removed. The sutures naturally decompose in the body.

Dissolvable stitches are usually made from natural material, such as processed collagen from an animal intestine. "Catgut" is a common name for this material, although it is actually made from sheep and beef intestines. The stitches dissolve over time, and by the time they do, the wound or incision has healed. Your body treats stitches as a foreign substance, and it is programmed to destroy them. Sometimes a stitch is left on the outside of the body and will not decompose because it is not in contact with body fluids. In these cases, the surgeon can remove any remaining pieces of suture material. Dissolvable sutures are available with a chromium salt coating to make them last longer in the body (chromic gut) or heat treated for more rapid absorption (fast gut). Surgeons choose one of these or untreated sutures (plain gut) depending on how long they expect the stitches will need to remain in place for complete wound healing.

A major portion of dissolvable sutures are now made from synthetic polymer fibers, similar to tennis racket strings or fishing line. Some are braided, others are monofilament. They are easier to handle, cost less, cause less tissue reaction, and are nontoxic. Many stitches are now made of cotton, where the wound is less prone to infection and the material is cheap.

Nonabsorbable sutures, made of materials that cannot dissolve in the body, are usually used for skin wound closures, because sutures on the surface are easy to remove. Sutures for this purpose are made from polyester, nylon, or silk. Stainless-steel sutures are used in orthopedic surgery and to close the chest after cardiac and abdominal surgery.

Stitches, or sutures, vary in thickness, elasticity, and decomposition rates. An extremely thin suture is used for cosmetic surgery, in which scarring is a major concern. An elastic suture is useful for knee surgery, so the knee can bend without tearing the stitches. A deep, wide gash needs stitches that will last longer, since it will take more time to heal than a simple surface wound.

130. How do huge ships float?

There was a time when very bright people thought that steel or iron ships couldn't float. After all, steel is about eight times as dense as water. Put a hunk of steel in water, and down it goes. It wasn't until the mid-1850s that steel was used extensively for shipbuilding. Prior to the Civil War, ships were made of wood. Everyone knew that wood floats in water.

Ships float because of their shape. A hunk of steel will sink. But that same hunk of steel can be shaped to make a lot of hollow space inside. When in water, the ship, with its outward-curving sides, is able to push out of the way, or displace, a weight of water equal to its own weight.

The *Titanic*, for example, weighed about 52,000 tons, so the floating *Titanic* displaced 52,000 tons of water. Put 10,000 tons of people and cargo on board, and the *Titanic* would displace 62,000 tons of water. Of course, then the *Titanic* would sink lower in the water. If the cargo weighs too much, or if a lot of water comes in the sides or over the top, bad things happen. Suddenly, the *Titanic* weighs more than the weight of the volume of water it can displace, and down it goes!

Archimedes, living in the third century BC, is credited with discovering the principle of buoyancy, which states that the upward force on a submerged object is equal to the weight of the fluid the object displaces. From this principle, he developed many related theories on water displacement. Famously, King Hiero of Syracuse commissioned Archimedes to determine if a gold crown was really made of pure gold or if the goldsmith had mixed in some other material, like silver. Sitting in the bath one day, Archimedes noticed that submerging his body caused the water level to rise. Archimedes reasoned that a submerged object displaces a volume of water equal to its own volume, so if the gold crown actually had silver mixed in it, it would take up more space than a pure gold crown because silver is less dense than gold. Archimedes compared the amount of water displaced by the crown and the amount of water displaced by an equal mass of pure gold. Sure enough, the crown displaced more water, indicating it was a fraud. The goldsmith literally lost his head. A very good read about Archimedes's experiment can be found at http://www.bellarmine.edu/eureka/about/.

131. What happens when an airplane gets struck by lightning?

Lightning strikes on aircraft are quite common, but also quite harmless. According to the US Federal Aviation Administration (FAA), on average, every airliner gets hit more than once a year. Of course, some planes get hit more than once and some not at all. Yes, indeed, that adds up to a lot of struck planes.

Modern planes have an aluminum "skin," and aluminum is a very good conductor of electricity. When lightning strikes a plane, it flows over the skin and off into the air. However, lightning could damage very sensitive electronic instruments on board, so airplanes require built-in lightning protection systems.

The last time an airplane crash was blamed on lightning was in 1963, when lightning struck a Boeing 707 that was in a holding pattern over Elkton, Maryland. The lightning spark ignited vapors in a fuel tank, causing an explosion and killing the eighty-one people on board. The very next week, on orders from the FAA, suppressors were required to be installed in aircraft flying within US airspace. These static dischargers are wicks or rods intended to drain off a charge of electricity in the event of a lightning strike. They are installed on the wing tips and horizontal and vertical stabilizers.

Pilots do try to avoid thunderstorms, not only because of lightning but also because the high shear winds (updrafts and downdrafts) can tear an aircraft apart. Also, if a jet engine pulls in a huge amount of water, the engine will quit (flame out). In addition, an airplane flying into a cloud that has built up an electric charge can actually trigger a lightning strike—but as explained, they can handle it.

132. How does a prism work?

Prisms are blocks of glass or plastic with flat polished sides arranged at precise angles to one another. Prisms can refract or deviate a beam of light and invert or rotate an image.

Prisms can also disperse, or break up, light into its various wavelengths. In 1666, a twenty-four-year-old Englishman named Isaac Newton blocked off all the light coming into his study, except for a narrow shaft of sunlight entering through a crack in the door. He used two prisms to disperse the sunlight and prove that so-called white light is composed of the seven colors red, orange, yellow, green, blue, indigo, and violet. We call these the seven rainbow colors. The mnemonic device ROY G BIV is used to remember the seven colors and their order in the rainbow (see page 48).

The separation of light into a spectrum, called dispersion, results from the refraction, or bending, of light. But the amount of refraction depends on the length of the light wave. Red, the longest light wave we can see, bends the least, whereas violet, the shortest wave our eye can detect, bends the most.

Binoculars use four prisms, two for each eye, to lengthen the path of the light coming through them and invert the image. We don't want to look through binoculars and have things appear upside down. With two lenses, the ocular, or eyepiece, and the objective lenses, any object appears inverted. Any land-use telescope, such as a spotting scope for wildlife, or a hunting scope mounted atop a rifle, employs a third lens, called an erecting lens. Having a third lens in the telescope necessarily makes the telescope quite long.

A long telescope is not practical for the user—it certainly could not be used with a neck strap. The prisms in binoculars reduce the light path and provides correct image orientation at the same time.

Periscopes in submarines and army tanks use two prisms. Both binocular and periscope prisms have three sides. They have three angles: Two of them are forty-five degrees and the other one is ninety degrees.

You can go to novelty shops and gift stores and buy prisms that hang in windows to create beautiful rainbows that dance around the interior of the house.

133. How do radar detectors work?

Radar was developed by the British prior to World War II. It was instrumental in giving the Brits advance warning of German fighter and bomber formations coming across the English Channel from occupied France. The name is an acronym for RAdio Detection And Ranging.

Radar detectors are radio receivers that pick up the radio transmissions from police radar transmitters. The concept of measuring vehicle speed by radar is really quite simple. The speed gun police use is a radio transmitter that sends out radio waves and picks up, or receives, any reflected radio waves.

The simplest use of radar is to tell how far away an object is. The radar transmitter sends out a concentrated radio wave. Any object in its path reflects a tiny portion of the wave back to the radar set to be picked up by a receiver. Since radio waves travel at a known speed, the speed of light, the radar set can calculate how far away the object is based on how long it took the radio wave to return.

The kind of radar used to measure the speed of a vehicle is known as Doppler radar. If a vehicle is approaching a radar unit, the radio wave travels a shorter distance in its round trip than if the vehicle were standing still; this has the effect of compressing or squeezing the wave together. That is, the wavelength shortens, so the frequency increases. Based on how much the frequency changes, a radar unit can calculate how fast the car is moving. Just the opposite is true for a vehicle moving away from a radar set. The radio beam has to travel a greater distance to reach the car and return. This has the effect of stretching out the wave, giving it a longer wavelength and lowering its frequency. If the radar unit is in a moving police car, then the speed of the police car must be factored in.

More and more police departments are using lidar, or LIght Detection And Ranging. This speed gun sends out short bursts of infrared laser light that bounce off a vehicle and return. The advantage is that the beam is much narrower than that used in conventional radar, so it's easier to target individual cars in heavy traffic.

Law enforcement often uses VASCAR (Visual Average Speed Computer And Recorder) to foil radar detectors. VASCAR is really

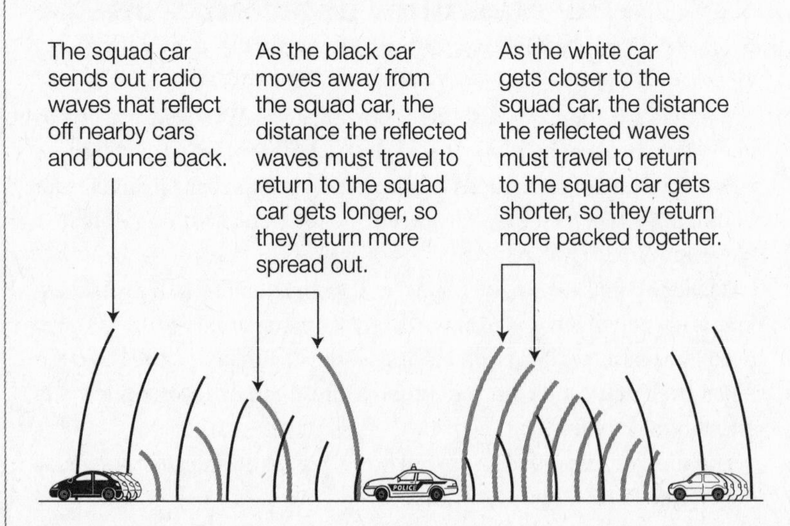

a stopwatch married to a calculator. Once a distance is entered, the operator pushes a button to start and stop a clock when a vehicle passes the first and second of two marked positions. VASCAR displays the speed based on distance and time. No radar is used. In Wisconsin, where I'm from, the State Patrol "bear in the air" employs VASCAR. The pilot clocks a car for the one-eighth of a mile between painted stripes on the Interstate then radios a waiting squad car on the ground.

I personally feel a real kinship to radar, as I have been issued six speeding tickets in about fifty years of driving. Deserved every one of them!

134. How do cell phones work?

A cell phone, or cellular telephone, is actually a two-way radio. Alexander Graham Bell invented the telephone in 1876, and Nikola Tesla and Guglielmo Marconi made their contributions to the invention of the radio in the 1880s and 1890s. So it seems like these two technologies were destined to be married.

Any area of cell phone coverage is divided into "cells" of about ten square miles, much like the hexagonal cells that bees store their honey in. Each cell has a tower and small transmitter building (base station). Each base station and every cell phone runs at low power, so the same frequency can be reused for nonadjacent cells. Across any densely populated area like a city, over eight hundred radio frequencies are available. In any one cell, about 160 people can talk on their phones at the same time. That's with the analog system. It's about 800 with digital. That may not sound like a lot, but remember that any one caller may be using that cell for just a few seconds, and then the signal is handed off to another cell or another tower. Some areas of the country still rely heavily on analog technology as part of their system. The system we are on in Tomah, Wisconsin, where I live, is all digital, which is much more power-efficient. In a digital area, my cell phone battery is good for about three days. Recently, I taught a Saturday class in the Milwaukee area. My battery was good for less than one day, because that system was part analog.

When it is first turned on, your cell phone listens for a system identification code (SID) on the control channel. If the cell phone can't find any control channels to listen to, it knows it is too far away and gives a "no service" indication. When it does receive the SID, the phone transmits a signal called a registration request, so the system's mobile telephone switching office (MTSO) knows what cell you are in. When you get a call, the system looks into its database to see what cell you are in. It then picks a pair of radio frequencies for you to use, one to listen to and one to talk on. You're now talking by two-way radio. When you get to the edge of your cell, the base station in that cell notes that your signal strength is decreasing. The base station in the next cell senses that your signal strength is increasing. The two base stations eventually "decide," using the control channel, to switch frequencies, and you are handed off to the next cell. It is so seamless that it is almost like a miracle, and yet you do not even notice it.

The circuitry in a cell phone is so complex that it would have taken up an entire room fifty years ago. Now it fits in the palm of the hand. Cell phones are becoming more sophisticated and versatile all the time; some play games, others contain PDAs, MP3 players, and cameras. In the United States, about 90 percent of the population uses a cell phone.

Whether it is used to summon life-saving help or simply to connect friends and family across the miles, the cell phone is truly an electronic marvel.

135. How does a lightbulb work?

Ah, the lowly lightbulb (see page 174). Little do we realize how Thomas Edison's 1879 invention completely transformed the way we live and work. Imagine how hard it must have been to illuminate our world once the Sun went down, using messy candles, torches, and oil lamps that tended to suck up the oxygen, soot up the room, and start disastrous fires. By 1900, millions of people around the world were turning on the lightbulb and turning off the dark. And amazingly, the lightbulb has not changed much in one hundred years.

The heart of the lightbulb is the tungsten filament. Tungsten is a metal that has an extremely high melting temperature of about 6,150°F.

But even tungsten would catch on fire at such high temperatures. Fires need oxygen, so bulb makers take out the air inside the bulb. In other words, there is a vacuum (nothing) inside the bulb. Then those clever bulb makers go one step further. They put a tinge of argon gas in the lightbulb. Argon is an inert gas and will prevent oxygen from the corroding the hot filament.

Bulbs burn out when too much tungsten has evaporated away in one particular spot on the filament, making the filament thinner and weaker in that area. The filament is cold, you turn on the bulb, a heavy surge of current goes through that cold, weakened filament, and, wham, a flash of light and the bulb is dead. We've all seen it.

The one-inch filament in a typical bulb is actually about twenty-two inches long. It is coiled, and then the coil is made into a larger coil. Engineers discovered they could get more light per watt with this double-coil method. Lightbulb manufacturing is a beautiful example of compromise engineering—a compromise between bulb life and bulb efficiency. You could make a bulb last practically forever: just make the filament really thick. But the bulb would be dim and consume lots of electricity. Not very efficient. On the other hand, you could make a bulb extremely efficient by making the filament thin. It would burn hot and bright but last just a few hours.

Every bulb package is required to have these items on the label: wattage, life in hours, light output in lumens, and estimated energy costs. A twenty-five-watt bulb will last about twenty-five hundred hours, and a hundred-watt bulb will last an average of seventeen hundred hours. The lower the wattage, the longer the life. That's because the filament is thicker on the low-wattage bulb. But the hundred-watt bulb is more efficient, which means you get more light per watt of electricity consumed. The newer CFL and LED bulbs may cost more, but are much more efficient.

CNN had a story concerning a lightbulb in a fire station in Livermore, California, that has been burning almost continuously since 1901. That bulb is not very bright and not very efficient, but it sure has lasted a long time.

136. How do lasers cut things?

Very powerful carbon dioxide (CO_2) lasers are used for cutting, welding, drilling, fusing metals (cladding), and surface treatment of metals. These CO_2 lasers, some up to ten thousand watts, emit light in the infrared and microwave regions of the spectrum. Infrared radiation is heat, so the CO_2 laser basically melts through whatever it is focused on. We cannot see the beam from a CO_2 laser, because its radiation is outside the visible part of the spectrum. The laser with waves are about twenty times as long as light waves that we can see. Other lasers, such as diode lasers, are very weak and are used in laser pointers (so their light is obviously in the visible spectrum), CD and DVD players, and bar code readers at the supermarket.

The Nd:YAG laser is a cousin to the carbon dioxide laser. Nd:YAG means neodymium-doped yttrium aluminum garnet. This is a solid-state laser that also emits light in the infrared region of the electromagnetic spectrum, but closer to visible light. The Nd:YAG laser has found a home in medicine, where it is used for eye operations and removal of skin cancers, hair, and veins on the face and legs. Solid state lasers do not have an evacuated tube or bulb.

What are the advantages of using lasers in industry? The beam can be aimed precisely, which minimizes heating of the surrounding material. The newer CO_2 lasers have low maintenance, use less power, have low operating costs, and take up less space. Many laser machines deliver the light via fiber optics, which makes it simpler to integrate the laser into workstations. Some unusual applications of lasers in manufacturing include cutting the Kevlar masks for hockey goalies, bonding cladding for surgical instruments, cutting the holes in baby bottle nipples, putting identifying marks on parts, and etching the plates used in color printing. Way back in the mid-1980s, I saw a laser weld the cases for heart pacemakers at Medtronic in Minneapolis.

Some low-power CO_2 lasers find uses in surgery. A small red laser guides the CO_2 beam. Retinal eye repair was one of the early medical uses of this laser. The laser is especially valuable in operations that tend to cause heavy bleeding, such as those on the liver. The beam instantly cauterizes the tissue, helping prevent infection.

Powerful lasers have been developed by the military as part of "Star Wars" research. Pictures and video have been released of lasers mounted on army tanks and in aircraft, but not much is known publicly about their power.

137. How do oxygen generators work?

Oxygen generators became big news when a ValuJet plane crashed into the Florida Everglades in 1996. Later, we heard about a fire aboard the Russian *Mir* space station. Both accidents were caused by oxygen-generating canisters.

When we think of storing oxygen, we think of those tall green cylinders at welding shops and hospitals, and smaller tanks that scuba divers take with them underwater. These big, heavy tanks hold oxygen in gas form. Their size and weight make them impractical as emergency oxygen supplies on aircraft, submarines, and space stations. There are alternatives to using tanks of oxygen: The umbrella term for them is "oxygen generator." Oxygen generators are the source of the emergency oxygen supply for airplane passengers in the event of cabin pressure loss. Submarine crews, firefighters, and mine rescue personnel also use them. Oxygen generators store oxygen in the form of a few different chemicals; sodium chlorate ($NaCLO_3$), barium peroxide (BaO_2), and potassium chlorate ($KClO_3$). Both are very rich in oxygen; canisters packed with them are small and lightweight, about the size of a tennis ball can. When airline passengers need supplemental oxygen, the canisters are released automatically by the drop in cabin pressure or manually. Pulling down on the face mask pulls out the retaining pin, and the chemicals combine to produce oxygen.

When activated, oxygen canisters become so hot that they can set anything combustible nearby on fire, and it will burn intensely because of the rich oxygen supply. In the May 11, 1996, ValuJet Flight 592 crash, five boxes of oxygen generators were stored in the forward cargo compartment. They did not have any safety caps on them. (They were listed on the manifest as empty and were to be sent back to Atlanta.) The investigators speculated that a jolt during takeoff ignited one of the canisters and in minutes the whole plane was awash in flames. The jet slammed into the Everglades swamp at 500 mph and at a steep angle. All 110 crew and passengers on board perished.

138. How do plasma television sets work?

Plasma television is relatively new and still fairly expensive. But finally, a flat-screen, hang-on-the-wall-like-a-picture television is here—if you can afford it!

What is plasma? It is the same stuff that is in our overhead fluorescent lights. Plasma is a gas that has a lot of uncharged free-flowing electrons (ions). Plasma is considered a fourth state of matter. The idea is to illuminate tiny colored fluorescent lights to form an image. Each pixel consists of three fluorescent lights, a red light, a green light, and a blue light.

With these three colored lights, we can get any conceivable color we can think of (see page 49). Each of these tiny cells actually looks like a little box in which the four walls and bottom, or floor, are coated with phosphors. The "box," or cell, is filled with either xenon or neon gas. Applying a voltage across the cell makes the free electrons in the cell collide with atoms of the gas and create ultraviolet light. The ultraviolet light causes the three different phosphors to glow. It is these glowing phosphors that make up the picture we see. The tiny gas-filled cells are sandwiched between two sheets of glass. Tiny electrodes run between the cells, delivering varying voltages that turn the cells on and off and also vary the intensity of light from each cell.

There are several advantages to plasma displays. There is, theoretically, no limit to the size of the screen. The screen can be very wide because, rather than being lit by a single projector, which would cause dimmer lighting in areas of the screen farthest from the projector, each cell is individually lit and the image is very bright from any viewing angle. The thinness of the display reduces its weight. Many of the big-screen displays at ballparks are plasma displays, as are advertising displays in malls and banks. The screen is bright over a large viewing angle.

139. What makes a refrigerator so cold?

Remember how cool your skin felt when you put a little rubbing alcohol on that last mosquito bite? The alcohol evaporated away quickly. It went from a liquid to a vapor, or gas. It takes heat to change a liquid into a gas, and that heat comes from our body. When you go swimming on a hot summer day, climb out of the water, and stand in the open air, you feel cool until you towel off. While you are standing there, water is evaporating from your body. It takes 540 calories to evaporate one gram of water, and those calories come from our body, causing a loss of body heat. Once your skin is dry, you start to feel hot again.

Modern refrigerators work the same way. The evaporation of a liquid carries heat away. A compressor condenses the refrigerant gas; compression causes any gas to heat up. The serpentine coiled tubing on the back of the refrigerator gets rid of this heat to the surrounding air as the gas condenses further and changes into a liquid. The higher-pressure liquid refrigerant goes through an expansion valve. The refrigerant immediately boils, i.e., goes from a liquid to a gas. Remember, it takes heat to change a liquid to a gas. The process draws that heat out of the food inside the fridge. Another is a set of serpentine evaporating coils inside the refrigerator (usually by the freezer compartment). The more coils, the more surface area for heat exchange. The compressor sucks up the cold gas, and the whole cycle starts over.

What a change in lifestyle the refrigerator has given us! The basic function of refrigeration is to slow down bacterial growth: It takes longer for food to spoil when it's cold. Milk will spoil in two or three hours if left on the dining room table. In a refrigerator, the milk will stay fresh for two or three weeks. Frozen milk can last for months.

Many industrial refrigeration units use ammonia. Ammonia is toxic and could kill people if leaked. In the 1930s, DuPont developed chlorofluorocarbons (CFCs) for use in home refrigerators. In the 1970s, it was found that CFCs damage the ozone layer, and they were phased out by an international treaty. R-22 was the refrigerant of choice for forty years, but R-22, like other hydrochlorofluorocarbons, also turned out to be bad for the ozone layer. The Clean Air Act prohibits the release of any R-22 when installing, servicing, or retiring older equipment. Now there are newer refrigerants that are safer for both people and the ozone layer. R-410A is an example of a hydrofluorocarbon that is friendlier to the ozone layer and widely used today in air conditioners and heat pumps.

140. How do counterfeit detector pens work?

A very good question. I went to a local office supply store and bought a counterfeit money detector pen made by MMF Industries so I could experiment with one myself. The instructions state that if the money bill is real, the pen leaves a light gold mark. If the bill is phony, it leaves a black mark. The iodine solution in the pen reacts with the starch in wood-based papers, leaving a bluish-black streak. Normal paper used for newspapers, construction paper, and typing paper come from trees, or from recycled paper originally from trees. The process breaks down the tree's wood into cellulose fibers, which are made into thin sheets of paper. But paper currency, on the other hand, is made from the same stuff rags are made of, mostly cotton and linen. There is no starch in our paper money. So the ink from the pen makes a light gold mark.

In addition, the fibers in our paper money bond together very tightly, and water does not get between them. It can go through the washing machine and come out intact. (Is that what they mean by money laundering?) Regular papers, made of loosely bound cellulose fibers, absorb water and tear apart when wet.

141. How does the metal wiring in a toaster heat up and stay hot?

A toaster, like so many things we use in everyday life, seems quite simple but really has some neat, useful, and incredible science principles behind it. The "secret" of the toaster is the nichrome wire that is wrapped around thin sheets of mica or aluminosilicate material. These materials are cheap and fireproof. You can actually see the red glowing nichrome wire when you peer down into the toaster. Nichrome wire is an alloy, a mixture of nickel and chromium. It's the same kind of wire that heats up hair dryers, clothes dryers, electric stoves, and electric space heaters.

Nichrome has two properties that make it a great producer of heat. The first is that nichrome has fairly high resistance to electrons. We know that everything is made up of atoms, the basic units of any element. Within the atom, even tinier negative electrons orbit a dense positive nucleus—much like planets orbiting the sun. An applied voltage is the driving force that gets electrons moving in a wire. The applied voltage is analogous to the water pressure in the water pipes in your house. Water pressure pushes the water through the pipes. In the case of toaster wires, it is the voltage that drives the electrons along on the wires. In nichrome, some of these electrons do not "behave." Instead of orbiting a single atom, they migrate through the wire and bump into and jostle other atoms, causing them to vibrate. That vibration and jostling heats up the wire, throwing off infrared radiation. Infrared radiation dries out the bread while charring the surface. Just add butter and jam!

The second unique property of nichrome wire is that it does not oxidize when heated. The tungsten filament in a lightbulb combines with oxygen, making it more brittle. Nichrome does not mix with oxygen to produce any change in its structure. That means it can get quite hot without wasting away and burning through. Also, nichrome doesn't rust easily, so you won't have iron-tasting bread!

How does the toaster know when the toast is done? The older method uses a bimetallic strip, which converts a heat change into mechanical movement. A bimetallic strip consists of two metal strips, typically steel and brass, bonded together. Most all metals expand when heated and contract when cooled, but not all metals do this at the same rate. Brass expands with heat more than steel, so as the strip heats up, it bends. Brass is on the outside of the racetrack, so to speak, and has a longer distance to go. Eventually the curving strip hits and releases a switch, and up pops the toast. Thermostats work much the same way, using a bimetallic strip to turn our furnace on and off.

Newer toasters use a simple circuit with a microchip, charging capacitor, and electromagnet. The capacitor is a device to store electrons.

The darkness control (light or dark toast) is a variable resistor. The resistor is a device that can limit the amount of electrical flow. The capacitor charges through the resistor, and then when the correct voltage is reached, the electromagnet is released and up pops the toast.

The toaster arose out of a need to make toast, which helps bread to last longer, without a burning fire, and the first commercially successful ones were made by General Electric in 1909. They had a wire heating element on a porcelain base, and people needed to unplug them or else the toast would burn.

142. Why don't we build more nuclear power plants?

The conditions may be right for a fresh look at nuclear power for generating electricity. With higher prices for natural gas and an uncertain oil supply from the Middle East, some people are rethinking nuclear.

First off, there is no free lunch. Every source or method of generating electrical power has advantages and disadvantages. While nuclear plants produce radioactive waste and pose a risk of meltdown, they produce enormous amounts of energy without creating greenhouse gas emissions.

We only have three methods of generating electricity in any appreciable quantity: hydroelectric (water), fossil fuels like coal and natural gas, and nuclear.

Hydropower, of course, is the most desirable, since, although it can harm the environment when dams flood areas upstream, it's cleaner and renewable. But all the good hydro sites, where rivers fall or flow some distance, already have generating stations. So further development of hydropower is not practical in the United States. However, finding new hydropower sites in Canada is a possibility. Some agreement as to shared development and use may be in the future.

We do have several hundred years' worth of coal in the United States, but coal comes at a huge environmental cost. Strip mining and underground mining can lead to a host of damaging results, including scarred landscapes, silt runoff, dynamite by-products, black lung disease, asbestosis, emphysema, mine collapses, and accidents between coal trains and cars.

Burning the stuff is even worse! Just one coal-fired power plant in one year produces, on average, roughly 3 million tons of CO_2 gases (which add to the greenhouse effect and global warming). Altogether, about three billion tons of CO_2 gases are released from our coal-burning electrical generating plants in just one year. Natural gas plants emit about 55 percent as much carbon dioxide as coal-operated plants. In addition to carbon dioxide emissions, coal plants release tons of sulfur dioxide and nitrogen oxides that cause acid rain, as well as mercury, which is toxic in the environment, including to the fish.

Why do we use coal? Because we have to, in order to provide the electricity that is so important for maintaining and increasing our standard of living, even as our population grows. Coal allows our factories to operate efficiently. Low-cost electricity is vital for industry to compete internationally and to provide jobs. But is coal our only choice? Advocates for nuclear power would say that, because of all the problems caused by coal, nuclear is definitely a viable alternative.

143. How do glow sticks, or light sticks, work?

Glow sticks work by a process called chemiluminescence, or the creation of light by chemicals. Because no heat is generated in the production of the light, this process is also known as cold light. The light sticks, or glow sticks, use a chemical called Cyalume, which was discovered to be luminescent by the German chemist H. O. Albrecht in 1928.

The basic process works like this: When an atom absorbs energy, some of its electrons move into an orbit with a higher energy level; we say they are in an excited state. The atom can release the energy as electromagnetic radiation, in the form of light, when the electron returns to an orbit with a lower energy level, which we call being in a ground state. In the case of the light stick, the triggering mechanism is the flexing of the light stick. A small glass vial filled with one chemical is inside a larger plastic cylindrical vessel filled with a different chemical. Flexing the plastic outer vessel breaks the inner glass vial, and the two chemicals mix.

Think of being on a bicycle at the bottom of a hill. You work hard to get to the top of the hill. You put energy into the effort. Now you're at the top of the hill. You have done work to get there. You can coast down the hill without any effort. If you were like an electron, then when you coasted down the hill, you would give off a flash of light. While most chemical reactions release heat, some release light. When Cyalume mixes with an oxidizing agent, usually hydrogen peroxide, a fluorescent dye accepts the released energy, converting it into light.

There are several recipes for light sticks. Most common commercial light sticks use a solution of hydrogen peroxide that is kept separate from a solution of a phenyl oxalate ester. A fluorescent dye coats the walls of the light stick. The decomposition of the two chemicals releases energy that excites the electrons in the fluorescent dye to give off light. You can now buy light sticks, or glow sticks, in various colors. I once mounted a red, a green, and a blue glow stick on each blade of a fan, and when the fan rotated, it created white light. The three primary colors for light are red, green, and blue. The combination of these three primary colors produces white light.

Most of the light sticks used for emergency lighting are green. Due to the chemical reactions involved, green is the easiest and cheapest color to produce. Green is right in the middle of the visible spectrum, the G in ROY G BIV (see pages 48–49). Our eyes are most sensitive to green light.

To do a fun experiment on your own, buy two green light sticks. Most hardware stores sell them, especially around Halloween trick-or-treating time. Bend the sticks so they glow green. Place one in warm water and one in cold water. Which do you think will give off light for a longer time? The cold-water light stick will last longer, because the low temperature slows down the chemical reaction. But you don't get something for nothing! The cold-water light stick will not give off as much light. The warm-water light stick will glow brighter but for a shorter period of time.

144. How do touch lamps work?

My gosh. How lazy we've become. Pretty soon you'll want a remote control for your TV and VCR! Scott M. Kunen invented the touch lamp in Freeport, New York, in 1985. His patent was updated in 1989 to include a box that plugs into the wall directly, with no additional wiring.

Several different properties of the human body make the various designs of touch lamps work, including body temperature. The body is usually warmer than the surrounding air and any nearby objects. So the warmth of the hand triggers an electronic switch operated by a thermocouple or thermistor device, an electronic element that changes its resistance with temperature. Many elevator buttons work on this principle. This does have the drawback that a fire can activate the button to call the elevator, taking your elevator to the floor where the fire is burning. Not a good idea! A good many of those motion-detector switches that control lights in our patio and garage areas sense the heat of the human body. Unless completely pure, water is typically filled with free ions like the salt in our bodies, which makes it a great conductor of electricity. Your finger can close the circuit between two metal contacts that are close together (see page 265). Many of the toys from fast-food places employ this principle. Most touch lamps also use the electrical capacitance of the body. As the name implies, every object has the capacity to hold electrons. The lamp itself can hold a certain number of electrons, but touching the lamp adds your capacity to the lamp's, and it takes more electrons to fill both you and the lamp. A circuit detects the change, and the light turns on. Those little plug-in modules that change any lamp into a touch lamp work the same way.

In a similar way, touching a radio antenna, especially the lower AM frequency as compared to the FM stations, will usually bring in a stronger signal. The human body is acting as an extension of the radio's antenna, bringing in a stronger signal.

Many touch lamps have three brightness settings. You get a higher brightness setting with each touch. A fourth touch turns the light off. The control unit rapidly turns the bulb on and off many times a second in what is termed the "duty cycle." The duty cycle is the amount of time the bulb is on or off. A duty cycle of 50 percent means the bulb is on half the time and off half the time. Variable-speed power tools work the same way. The voltage to the motor of a variable-speed drill in the United States is always 120 volts AC. But the hand trigger sets the time in which the voltage is delivered to the motor. In this manner, the high voltage of 120 volts maintains a high torque, or turning force, but delivering that voltage for a fraction of the time determines the speed of the drill rotation.

The advantage of touch lamps is convenience. You can brush the metal lamp stand with your hands, arms, or even your bare feet, if so inclined. Very handy if your arms are full. This type of switch also prevents unwanted materials from accumulating, the way they do in normal, old-fashioned mechanical switches. Dirt and debris can allow an easy path for the electricity to jump across contacts, causing a spark and wearing out the switch.

5

Stuff I Always Wondered About

145. What exactly is science?

The word "science" comes from the Latin word "*scientia*," meaning "knowledge," and is defined as "knowledge of something acquired by study." Science is the study of the world around you. Science explains the world we live in, how things work, how living things come to be, and how things happen the way they do, making it one of the basic human pursuits in life, much like music, art, and literature.

In the time of Aristotle (384–322 BC), science was linked with philosophy and the two words were used interchangeably. In the time of Isaac Newton (1642–1727), science was a branch of philosophy referred to as natural philosophy.

There are two main branches of science: natural science and social science. I am focusing on the natural sciences in this answer. Biology is the study of living things. Physics explains how the world works; the interaction of matter and energy, forces, motion, and mass. Chemistry explains how the universe is put together; the atoms, the elements, the molecules, and the interaction of all the chemicals. Earth science encompasses climate, weather, geology, astronomy, oceanography, and geomorphology.

People study the natural sciences so they can understand planet Earth and the larger universe better. For example, knowing how the atmosphere works may allow us to address global warming. Knowing about how the body works, with its cells and systems and life functions, has led to prolonging life and creating vaccines and cures for diseases. Knowing how the Internet works provides us a pathway and access to an unlimited amount of information. Each of these main topics of study includes a host of specialized fields. If you google the phrase "list of sciences," http://phrontistery.info/sciences.html, you find that 633 science titles pop up. The list ranges from acarology (the study of mites) to zymurgy (the chemistry of fermentation, usually brewing and distilling).

Everything around us concerns science and scientific processes, and we experiment, research, and follow the scientific method to discover these mechanisms of life. The scientific method starts with observation, moves to a hypothesis (educated guess based on that observation), and ends with a prediction or a theory. A theory summarizes a hypothesis that has been supported by repeated testing, and it remains valid as long as there is no evidence to dispute it.

Science fosters logical and critical thinking in order to figure things out. Science makes our lives much better, easier, and longer. Science can make life more enjoyable and pleasant and the world a safer place to exist through construction, navigation, transportation, agriculture, and medicine. Of course, one can argue that there is a downside to science. We humans have developed weapons that can destroy each other very quickly and in huge numbers. Sometimes it seems our knowledge of how to get along with each other lags behind our knowledge of how to destroy each other.

Science can make you less gullible. There is a lot of wrong or misleading information on television, and the Internet, in newspapers, and magazines and from politicians and government leaders and also from the private sector. Science helps us filter through the claims of advertising and helps us spot the hoaxes, scams, schemes, and cons that are out there. And, finally, science is a lot of fun. It can be bizarre, bewildering, exciting, and satisfying. Science complements literature, art, music, and religion. While religion allows us to believe in things that are not provable, science is an avenue to prove things that are hard to believe.

146. Why are some people smarter than others?

First of all, what exactly does it mean to be "smart" or have a high IQ? "Smart" is a term often used to describe academic excellence. We sometimes use "smart" as a synonym for "intelligent." But there can be a difference between academic excellence and intellectual potential.

While most students with high or above-average intellectual skills thrive in schools, many others with equal IQs seem lost in school. This is because there are many different kinds of intelligence. Howard Gardner, a Harvard developmental psychologist, developed a theory of multiple intelligences, including linguistic, logical-mathematical, musical, visual-spatial, bodily-kinesthetic, interpersonal, and intrapersonal. Gardner's landmark 1983 book, *Frames of Mind: The Theory of Multiple Intelligences*, proposes that we all possess all these intelligences, but each of us has them developed to different degrees and uses them differently.

The "nature vs. nurture" question has been debated for close to 150 years. How much of our "smarts" comes from our DNA—our parents' genes (nature)—and how much comes from the way we were raised or from our environment (nurture)? Which has more influence on our intelligence? We know that genes hold some degree of influence, but no one knows how much.

A study of forty thousand Dutch children showed that firstborns had a slightly higher IQ than their siblings. One theory is that firstborns tend to get more attention from Mom and Dad, and with more nurturing, they have a better foundation for academic achievement. How children are raised falls in the "nurture" aspect of the nature versus nurture discussion.

An individual's interests and work ethic can influence intelligence, too. Hardworking people tend to learn more than even very bright-minded but lazy people. And being passionate about a specific topic, say music or art or science, pushes a person's thirst for knowledge.

There are more factors. Can intelligence depend on wealth and social status? Yes, to some extent. A child who has been taught the importance of education will have a higher desire and motivation to learn than a child who hasn't. Also, people tend to marry other people with similar kinds of intelligence. Wealth matters, too. Poor families are unable to send a bright kid to college, while wealthy families not only put their kids through college but also hire tutors.

The brain is an organ, but treating it like a muscle can improve its function. The more you exercise it, the better and stronger and more efficient it becomes. So people can become "smarter" and raise their IQ simply by exercising their brain. Conversely, an idle mind tends to lose its abilities. That is why music for young kids is very helpful, and experts recommend reading, writing, crossword puzzles, etc., for senior citizens.

College-bound students in the Midwest take the ACT test. Students along our coasts tend to write for the SAT. An IQ test determines a person's general problem-solving ability and concept comprehension. It consists of tests of memory, spatial, logic, and math abilities. The companies that write the ACT and SAT tests do not bill them as IQ tests, but they are considered one of the best predictors of success in college. That's why admissions people look primarily at ACT and SAT scores and class rank in determining who gets into colleges and universities.

Last, please keep in mind that being "smart" or having a high IQ is just one aspect of a person's being. An IQ test does not measure creativity, empathy, kindness, or even motivation. Talent can't be measured accurately by IQ tests, especially in the areas of music, art, dance, writing, social skills, or, most important, people skills.

147. Why do people in some countries drive on the left side of the road?

In almost all countries of the world where people drive on the right-hand side of the road, the cars are made so that the driver sits on the left-hand side of the car. That's the case in the United States (with the exception of the US Virgin Islands). The driver sits on the side of the car that is nearest the centerline of the highway.

Most of the world, in 165 countries, drives on the right side of the road. The exception: seventy-five countries drive on the left side. The most notable countries that drive on the left side of the highway are the United Kingdom (England, Scotland, Wales, Northern Ireland), Ireland, most of India, and Australia. You might recognize that countries that were part of the 1800s and early 1900s British Empire tend to drive on the left side.

The reasons are historic. In olden days, people who passed each other wanted to be in the best possible position to protect themselves. Since most people are right-handed, they therefore kept to the left. Their useful hand would be next to the person passing them. Also, mounting a horse on the left, which is easier for right-handed people, avoided accidents stemming from the sword getting between the horse and rider. The sword was typically worn on one's left, so that it could be pulled out of its scabbard from across the body with the right hand.

It is widely believed that Pope Boniface made left-side travel an edict in the 1300s. The United Kingdom made it a law in 1773, and it became part of the Highway Bill of 1835. The reason behind France's right-handed travel intertwines with the French Revolution beginning in

1789. The French aristocracy drove their carriages at high speed on the left side, forcing the peasants, who had to walk, to keep to the right-hand side. During the revolution, the French peasants began lopping off the heads of the French royals at a pretty good clip (no pun intended), so for self-preservation, the royals slowed down, kept a low profile, and blended in with the crowds on the right side.

The first official record to keep right was started in Paris in 1792. France also built up quite an empire in the early 1800s, with Napoleon leading the way. Countries conquered by the French, such as the Low Countries—Germany, Poland, Spain, Italy, and Switzerland—drove on the right, while the British Empire people continued to drive on the left.

It is thought that the French connection to the United States came by way of General Lafayette, who helped us greatly in the Revolutionary War. A law governing Pennsylvania's Lancaster and Philadelphia Turnpike in 1795 is the first reference to driving on the right in the United States.

Canada is an interesting case. Provinces controlled by the French, such as Quebec, drove on the right, while British Columbia, New Brunswick, Nova Scotia, Prince Edward Island, and Newfoundland drove on the left. British Columbia, on the Pacific coast, and the Atlantic provinces switched in the 1920s. The rest of Canada switched to the right side after World War II. Now all Canadians drive on the right side of the road.

148. Why did the Twin Towers fall if planes only crashed into the top floors?

The National Institute of Standards and Technology (NIST) was mandated to investigate the September 11, 2001, collapse of the twin World Trade Center towers as well as the forty-seven-story 7 World Trade Center building. Their final report was released on October 26, 2005. According to their website, the NIST concluded that the WTC towers collapsed because "the impact of the planes severed and damaged support columns, dislodged fireproofing insulation coating the steel floor trusses and steel columns, and widely dispersed jet fuel over multiple floors . . . [T]he subsequent unusually large number of jet-fuel ignited multi-floor fires (which reached temperatures as high as 1,000 degrees Celsius, or 1,800 degrees Fahrenheit) significantly weakened the floors and columns with dislodged fireproofing to the point where floors sagged and pulled inward on the perimeter columns. This led to the inward bowing of the perimeter columns and failure of the south face of WTC 1 and the east face of WTC 2, initiating the collapse of each of the towers."

Video evidence shows that both towers, each weakened on one side, toppled slightly, with top floors collapsing first and the lower floors pancaking one on top of the other from top to bottom, until they disappeared in a cloud of dust and smoke.

There are numerous conspiracy theories in print and on television about the collapse of the WTC towers. The most prevalent one is that someone placed thermite and shaped charges on the supporting columns near the base of the towers. A shaped charge is an explosive charge shaped to cut metal. Thermite is an incendiary material used by the military to cut through steel. It is also employed by demolition experts to bring down old buildings and is often used to join rail tracks. Thermite, a mixture of iron oxide and aluminum powder, releases intense heat, up to 5,000°F. There were eyewitness accounts of molten aluminum in the rubble pile, which continued to smolder well into December. These accounts are probably true, but any molten steel in the wreckage was due to the high temperature resulting from long exposure to combustion within the pile rather than short exposure to fires or explosions while the buildings were standing.

Could thermite have been responsible for bringing down the towers, as some have proposed? Estimates are that thousands of pounds of thermite would have had to be placed inconspicuously ahead of time, held in contact with the surface of hundreds of massive structural beams, and then remotely ignited to weaken the building. And thermite burns rather slowly, so these burns would have to be coordinated with the time the planes struck the towers. That's a tall order, even for very smart terrorists. Steel loses half its strength at 1,200°F and melts at about 2,800°F. So the fires in the upper floors, burning with over ten thousand gallons of jet fuel, did not have to be more than 600 to 800°F to melt the insulation surrounding the steel beams, weakening them sufficiently to cause the weight of the floors above them to collapse.

NIST used some two hundred technical experts, reviewed tens of thousands of document pages, analyzed 236 pieces of steel from the wreckage, performed hundreds of lab tests, and used sophisticated computer simulations of the sequence of events from the moment the planes struck until the towers collapsed.

NOVA on PBS and the March 2005 issue of *Popular Mechanics* did a good job of explaining the Twin Towers' collapse and dealing with most of the conspiracy theories.

149. How are coins made?

The very first coins (usually made of gold or silver) were minted in the area of modern Turkey about 640 BC. The custom spread to the Greeks and then to the Romans.

The US Mint, headquartered in Washington, DC, makes coins in Philadelphia, Denver, and San Francisco and has facilities at West Point and Fort Knox. Coin making is a long, complex process, going from raw material metals to finished coins. Modern US coins have their beginnings in the private sector, where a number of companies produce some coinage blanks, called planchets, and strip metal that the mint purchases. The mint produces its own dies to minimize the risk of counterfeiting. A coin die is one of two metallic pieces that are used to strike on the side of a coin. The die contains the inverse version of the image to be struck on the coin. Coins are stamped out on a big machine, polished, and inspected. To view a very good slide show and video on coin making, go to this website: http://www.usmint.gov/faqs/circulating_coins/index.cfm?action=coins.

Coins for circulation are made in Denver and Philadelphia with the identifying D or P on the coin. Collector coins and sets are made in San Francisco. The mint makes about seventy million circulating coins a day, in six denominations.

According to the US Mint, it takes only a few cents to make a quarter. But the quarter is instantly worth a quarter, so the difference is a profit for the mint. But now, because of increases in the price of metals, the cost of making and circulating a penny is more than the value of the coin. The same is true of the nickel. So the mint loses money when it makes, distributes, and circulates a penny or a nickel coin.

We tend to think that a penny is made of copper, because the outside of a penny is copper-colored. But the copper is just a thin coating. Since 1982, over 97 percent of a penny is zinc. Zinc prices went up 73 percent in one year alone. The last time a penny was made of pure copper was 1856. At the present time, the US dime and quarter are approximately 92 percent copper and 8 percent nickel. The US nickel is 75 percent copper and 25 percent nickel.

Starting in 2009, the Lincoln Memorial was no longer on the one-cent piece. The US Mint issued four different one-cent coins commemorating the hundredth anniversary of Lincoln's face on the coin and the two hundredth anniversary of his birth. Each coin shows a different aspect of the life of Abraham Lincoln. In 2010, the penny got a new permanent design. Lincoln's mug is still there, but his memorial isn't. The composition is the same as before: copper-plated zinc.

Current law mandates the following inscriptions appear on all circulating coins: "Liberty," "In God We Trust," "United States of America," and "*E Pluribus Unum*," as well as each coin's denomination and the year it was minted.

150. **What is that watery haze above the road in the summertime?**

That's a mirage. It is an actual optical occurrence—not merely an illusion—and shows up in photographs. Mirage comes from the Latin "*mirare*" meaning to look at or to wonder at.

The "puddle of water" on the highway ahead is a common mirage caused by the refraction, or bending, of light rays. As the Sun beats down on the highway, the black surface absorbs heat, the warm air rises and forms a layer of hot air a foot or so deep above the road. The heated roadway keeps the air just above the road several degrees higher than the surrounding air temperature. Some of that warm air is trapped under the cooler air above. A nonuniform medium, or mini-inversion, has been created by the heating of the highway and the air just above it.

Light waves travel in a straight line in a uniform medium, but they refract or bend in a medium whose density varies. Light travels faster through less-dense warm air. Any light ray from the distant sky would normally hit the ground. But in this case, the light actually bends in midair because it moves from the cooler, denser air into the hotter, less-dense air right above the ground. The light speeds up in the less dense air, which refracts it upward to your eye.

The puddle of water we see on the road ahead is actually a piece of the sky. The hot road image is the most common type of what is known as an inferior image, that is, one that appears to be lower than the real object. Mirages in the desert occur the same way. Light is refracted when it travels through different densities of air or water. But if the angle at which light strikes any surface, such as water or glass, is too large, total reflection takes place.

The angle at which light is totally reflected is called the critical angle, and it causes the surface of water or glass to act like a mirror. Hot air reflects light from the sky up to our eye, which is why we don't see the mirage or puddle right in front of us but farther ahead. In fact, while the mirage may look like a puddle, what we are actually seeing is light from the sky. If you look carefully, you can sometimes see a fluffy white cumulus cloud in the puddle. If we look out our car windshield, we can see the clouds in the sky ahead of us. But we also seem

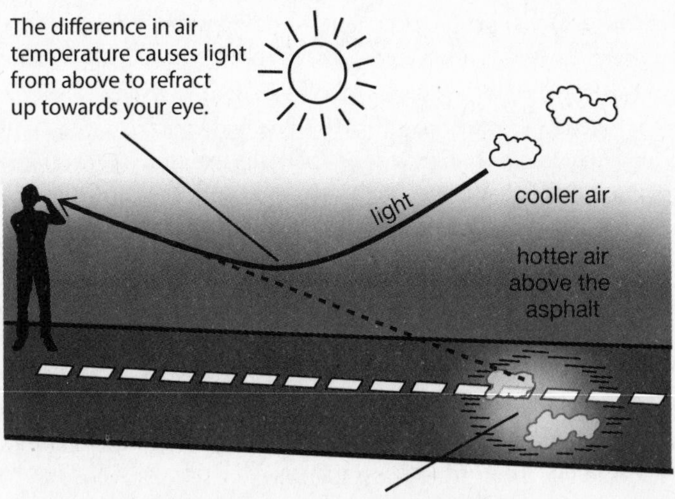

to see pieces of that same cloud in the roadway ahead. We can mentally trace a single light ray from the cloud. The ray travels at an angle to the highway ahead of our car, hits the less dense air above the roadbed, travels nearly horizontally for a bit, then bends upward to our driver eye. The bending is continual and gradual because the air temperature changes are gradual.

151. How can the exhaust from a car harm us?

Like so many things in life, owning and driving a car is a double-edged sword. Cars are powerful, convenient, and attractive and represent the freedom of the open road.

But cars are killers, both in accidents and in the pollution they emit. It has been said that driving a car is the most polluting act the average citizen commits. And emissions from cars are increasing, for two reasons: more vehicles and bigger vehicles. People are driving more vans, pickup trucks, and sport utility vehicles, although that trend might not last if gas prices keep increasing.

Here is a partial list of the pollutants found in car exhausts: carbon monoxide, nitrogen oxides, sulfur dioxide, benzene, formaldehyde, and suspended particles. Probably the most dangerous stuff in the exhaust is carbon monoxide. The hemoglobin in our blood transports oxygen from our lungs to the rest of our body, such as to the muscles, where it releases the oxygen for the cells to use. But carbon monoxide attaches to the hemoglobin much better than oxygen—over two hundred times better, and that keeps oxygen from attaching. So carbon monoxide poisoning is a form of suffocation. It puts a heavy burden on the cardiovascular system (see page 232). So it's very dangerous to run a car in a closed area, such as the garage. And, if possible, stand upwind from a car's exhaust. Most important, don't put your mouth over the exhaust pipe!

Benzene is a known carcinogen, or cancer-causing agent. Benzene affects our bone marrow, which produces red blood cells, and has been related to anemia and leukemia, and lymphoma. By the way, benzene is also found in cigarette smoke, solvents, and pesticides. It's the double-edged sword again, having both benefits, since it's an ingredient in some of the things we need, and negative effects.

Most cars made after 1996 have a standard port that allows a computer to diagnose engine performance. Computers have been added to engines to handle airflow in, fuel delivery, and exhaust out. The future does look bright. More hybrid cars, with much smaller engines, are being produced. Using ethanol may help some, too, since it produces less greenhouse gas than gasoline, although ethanol exhaust has more formaldehyde and acetaldehyde, which also may cause cancer, and arguments can be made that producing ethanol increases pollution, especially carbon dioxide, which contributes to global warming.

But try to stay away from car and truck exhaust as much as possible or practical. We can't help but inhale some of that stuff. Personally, I've taken to jogging on the path in the marsh area or in the cemetery instead of going around Lake Tomah, near my home. There are a lot of cars going by the lake on the roads during late-afternoon commute times. But people don't seem to bother you in the cemetery!

152. Why doesn't soap taste as good as it smells?

We're talking two different senses and two different sets of ingredients. Soap tastes bad because of the fats, oils, and caustic soda used in modern soap making. Soap smells nice due to the perfumes the manufacturers add.

The earliest soaps were made by boiling animal fat (tallow) or vegetable oil along with wood ashes. Soap didn't "take off" until after the Middle Ages. People simply did not bathe, believing that bathing would leave the pores of the skin open to diseases. Little wonder Henry VIII had six wives! What woman wants a husband who does not bathe?

The big breakthroughs in soap making came in 1779 when a Swedish chemist, Carl Wilhelm Scheele, produced glycerin, or glycerol, and in 1813, when the French chemist Michel Chevreul published the first of his many papers on the chemical composition of fats and oils. Glycerin reduces the harshness of soap on the skin.

"Saponification" is the fancy term used to describe the reaction of an alkali base with a fat to form soap. Sodium hydroxide (NaOH), also known as caustic soda or lye, makes hard soap, and potassium hydroxide (KOH) makes soft soap. Before World War II, soap making involved mixing fats and oils in large kettles, and adding one of these bases along with steam and some salt. The soap would precipitate out and float to the top, where workers would skim it off and make it into flakes or cakes. The process of soap making today is much the same, except higher temperatures and pressures and use of centrifuges have sped it up. The manufacturers add fragrances to make it smell nice.

Floating soap can be traced to an old legend, possibly untrue, from 1878, when a worker at the Procter & Gamble factory in Cincinnati, Ohio, supposedly forgot to turn off the soap-making machine when he went to lunch. When he came back, he discovered that too much air had been whipped into the batch and it was puffed up and frothy. He didn't want to admit his mistake, so he sent the batch down the line. It hardened and was chopped into cakes and sent out to be sold. People started asking for more of that "soap that floats." A new product was born, and in 1891, Procter & Gamble started printing the slogan "It Floats" on every label.

The floating soap story may be myth. According to Procter & Gamble archivist, Ed Rider, one of the company chemists, James N. Gamble, son of the founder, knew how to make floating soap as early as 1863.

When I was a little tyke on the farm, we made our own soap. My folks bought cans of lye from the store and we mixed it with the renderings of lard, or fat from hogs. We boiled the liquid and then poured it into pans to cool and harden. We used it for laundry soap, not for bathing. That stuff was so strong, it could take your skin off!

153. Why don't school buses have seat belts?

The National Highway Traffic Safety Administration has stated that seat belts are not the most effective way to protect schoolchildren. Instead, they subscribe to the idea of "compartmentalization," which means sitting in strong, closely spaced seats that have energy-absorbing, high seat backs and having the seats strongly anchored to the floor and frame of the bus. The primary argument against adding seat belts is that they would provide very little added protection for head-on impacts. Indeed, studies show that adding seat belts can actually increase the chance of head and neck injuries.

The challenge is that school buses carry passengers from the very young up to high school students. If seat belts are worn, they have to be properly adjusted for the size of each child. The seat belts must be snug and on the lower hips/upper thighs. A seat belt not properly worn can cause serious injuries. Lap belts worn without shoulder belts have proved to be unsafe for very small children, the kids in kindergarten or lower grades.

There are a number of other reasons school buses are not equipped with seat belts. First, school buses are big and heavy, kind of like a Mack truck or army tank. In a collision with a car, they seem to come out okay. Second, the cost of installing seat belts is high and seating capacity is reduced. Third, the fatality rate in school buses is already extremely low. Most school bus fatalities occur when a car strikes a kid getting on or off the school bus, and buses already bristle with safety features, including strobe lights, reflectors, and mirrors, all designed to address this leading cause of death. Last, if an accident did occur, such as a fire or an overturned bus, it would take a long time to unbuckle children and get them out.

A lot of the safety issues for school buses have had to do with making sure the driver is safe and qualified. School districts generally spend a lot of time, money, and effort to make sure that bus drivers are highly trained. The days of hiring a person "off the street" and putting them behind the wheel of a school bus are long gone. A commercial driver's license with school bus endorsement is required. The exact requirements vary from state to state, but the training is generally more than ten hours behind the wheel, with extensive pre-trip interaction, written tests, and a driving skills test with an examiner. And school bus drivers, like truck drivers, are subject to random drug testing.

154. How do you salt peanuts in the shell?

Salt permeates peanut shells when the peanuts are cooked in a briny solution under pressure for ten to twenty minutes. Brine is a mixture of salt and water, and the pressure forces the salt into the peanut shell. The excess water is removed by roasting the peanuts in a very dry oven that has a temperature of over 800°F. Some processors use a vacuum to remove air from inside the peanut shell before they are put into the brine. The vacuum inside the shell tends to draw the brine water into the shell. The roasting process changes the flavor of the peanut as the reaction of peanut sugars with amino acids forms flavor compounds.

What about peanuts that are already shelled? You and I might buy a bag of peanuts and shuck, or shell, them to get the peanuts out. But processors can't afford the labor costs. So they run batches of peanuts over screens and through blowers that remove leaves, twigs, and bugs and across magnets to remove metal fragments. Next, the peanuts are sized by rolling them over a series of screens with different-size holes. The smaller peanuts fall through the first ones, while the larger ones travel on to the next set of screens. Sizing is important in the shelling process, which involves rolling them between large metal drums with a certain gap between them. The gap is small enough to crack open the shell but big enough to let the peanut drop down without being crushed.

The peanut is an odd plant, because it flowers aboveground but its fruit is belowground. The peanut is a legume, same as alfalfa, clover, peas, and beans, and hence a nitrogen-fixing plant. Nodules that contain nitrogen-fixing bacteria form on the roots of legumes. The nodules absorb nitrogen from the air, and the bacteria enrich the soil with it. Then the legumes use the nitrogen to make proteins for themselves, which we benefit from when we eat these plants.

Peanuts prefer a warm climate, such as that found in the southern United States and in most of South America and Africa. Georgia grows more than 40 percent of all peanuts in the United States, with Texas, Virginia, the Carolinas, Alabama, and Florida producing most of the rest. Half of the peanut crop ends up in peanut butter.

155. What makes a golf ball curve?

The dimples in golf balls trap a thin layer of air next to the ball. This "boundary layer" spins with the ball. A golf ball hit off the tee has a tremendous backspin. Since the ball is traveling forward and with backspin, the layer of air on the bottom of the ball is traveling "against the wind," whereas the top layer of air is going "with the wind." The air is traveling over the top of the ball faster than is it is traveling under the bottom. The difference in air speed gives the ball more air pressure (a Bernoulli effect) underneath it (see page 152). Hence, the ball has lift, much like an airplane wing.

If a golfer does not hit the ball squarely, it may hook to the left or slice to the right (from a right-handed player's point of view). The ball is not only getting lift upward but also getting some unwanted lift to the left or to the right. To eliminate a hook or slice, the flat face of the club must be exactly perpendicular to the direction of the intended flight of the ball. If it is not, the ball will have a bit of clockwise (slice) or counter-clockwise (hook) spin.

I'll end with a golf joke: George and Ralph are starting to putt on the seventh green. A funeral procession is slowly passing by on a nearby road. George pauses, faces the passing hearse, stands at attention, removes his cap, and bows his head. A minute later, Ralph says, "George, that was an impressive thing you just did. I've never seen you touched so deeply." George replies, "Well, after all, we were married thirty-five years."

156. Why are tennis balls fuzzy?

I asked a few local tennis players, but their answers were also a little fuzzy, so I turned to several other sources. Professor Howard Brody has published articles in *The Physics Teacher* magazine and the *American Journal of Physics*, where he discusses the impact of the tennis ball with the ground. The fuzziness increases a tennis ball's contact with the ground. This raises friction, so the fuzzy ball is more likely than a smooth ball to roll rather than slide when it strikes the court. You might have noticed that tennis balls are cut open and attached to the legs of walkers as a safety device to prevent skidding or sliding.

All the skill shots of the game—topspin, backspin, and even sidespin—are more pronounced because of the fuzzy ball's gripping the ground. This means that a fuzzy ball will play differently on different surfaces, such as grass, clay, or rubberized concrete. A smooth ball would react pretty much the same way no matter what the surface.

One article, by Professor Brody, pointed out that fuzzy tennis balls may be based on tradition. Tennis has been around for over eight hundred years before the first fuzzy tennis balls were used, and the first tennis balls probably had cloth covers that got "fuzzier" the more they were used. Today, the fuzz is synthetic felt or wool surrounding a hollow ball that is pressurized to about twenty-seven pounds per square inch, which is about twice normal atmospheric pressure.

The tennis ball is a yellow-green color (called optic yellow) because the eye is most sensitive to yellow-green. Same holds for traffic signs around schools, many emergency vehicles, and neon-yellow softballs. TV sportscasters love the optic-yellow tennis ball, as it shows up nicely on television.

157. How many people live on planet Earth?

According to the US Census Bureau, world population is a bit over seven billion. The following list shows when each billion milestone was reached.

1 billion reached in 1802
2 billion reached in 1927
3 billion reached in 1961
4 billion reached in 1974
5 billion reached in 1987
6 billion reached in 1999
7 billion reached in 2012

The growth rate has dramatically slowed over time, largely due to decreasing birth rates. In the United States, there is a birth every eight seconds and a death every twelve seconds. The country also gains one immigrant every forty-four seconds. So there is a net gain of one person every thirteen seconds in the United States, or about 2.5 million people per year. When I was in high school, the population of the United States was 180 million. Today, it is approaching 320 million.

How many people can the world sustain? Best estimates, from a number of sources, put the number at about ten billion people. Eminent Harvard University sociobiologist Edward O. Wilson bases his ten billion estimate on calculations of the Earth's available resources.

The problem is not a lack of food, water, housing, medicine, or jobs. Rather, it is the distribution of these essentials that is problematic. We in the United States make up about 5 percent of the world's population, but we consume about 20 percent of the world's energy, for example. We live "high off the hog!"

Thomas Malthus's 1798 *Essay on the Principle of Population* observed that in nature, plants and animals produce far more offspring than can survive. He suggested that the human race, too, is capable of excessively reproducing if left unchecked. Malthus proposed that family size be regulated. People shouldn't have more children than they can support. Two centuries later, China is doing just that, with their policy of one child per family.

Paul Ehrlich argued that population growth would outstrip food production. His book *The Population Bomb* (1968) was a big hit when it came out. Ehrlich and other doomsayers are constantly predicting catastrophe in a global war as we fight for precious resources. They also believe that, eventually, the land will be stripped bare. Now we hear predictions that the lower birth rate in Western countries will lead to economic stagnation and decay.

There is a population clock at this website: https://www.census.gov/popclock/?intcmp=home_pop.

Personally, when I am looking for a parking space, I think there are too many people. And when I am in line at McDonald's, there are way too many people!

158. How risky is driving a car?

Some risks, like driving a car, are easy to determine. Take the number of people killed or injured in traffic accidents involving a car each year in the United States (about thirty-three thousand) and divide it into the number of people in the United States (about 320 million). So the odds of any American being killed in a car accident each year (granted, any particular individual's odds would depend on age, location, and a host of other factors) is about one in a little less than ten thousand.

Other risks are very difficult to quantify. What is our chance of being killed by a nuclear power plant? That's hard to determine, because no one has ever conclusively been proven to have been killed by nuclear power in the United States.

We can reduce our risks of being killed or dying of unnatural causes by careful monitoring of our lifestyle and behavior. People are aware of risky behavior, such as smoking, excessive drinking, overeating, not wearing seat belts, etc., but some engage in them nonetheless, often leading to addiction and disastrous results. We humans don't always use common sense or gauge risk scientifically. Some people refuse to fly in airplanes, for example, although air travel is, by far, the safest way to get from here to there. Lightning kills about fifty-three people per year in the United States, yet it's likely that a lot of those people were a tad beyond careless—like holding a metal rod up in the air (playing golf) during a storm.

Some actions can change the nature of risks. For example, if you decide it is safer to live in an underground house so that a car or plane won't hit it or a tornado or hurricane won't destroy it, then you increase the risk of death by other means. Your chances of cancer by radon gas or drowning or cave-ins can increase tremendously. You can only reduce your risk so much.

What's a person to do? Don't engage in known risky behavior. Observe Aristotle's Golden Mean for a successful and happy life: moderation in everything. And allow me to add: Enjoy the simple things in life!

159. Why is "pound" abbreviated "lb."?

The "lb." for a unit of weight is derived from the Latin words "*libra pondo*."

"*Libra*" means "scale" or "balance"—same as the astrological constellation in the heavens, Libra, a grouping of stars that resembles a scale with two pans in balance. The pound derives from the Latin "*pondo*," which translates to "by weight" and is related to "ponderous."

Furthermore, the ounce is abbreviated "oz.," which stands for the Italian "*onza*," which comes from the Latin "*unicia*," which means one-twelfth. At one time there were twelve ounces to the pound, not the sixteen that we use today.

The troy system, used by jewelers and dealers of precious metals, still has twelve ounces to the pound. Oh, it can be very confusing, which is why sensible people, and nearly all nations, with the notable exception of the United States, have all gone metric.

Science is filled with all sorts of symbols that may seem a little weird until you do some investigation. For example, the symbol for lead is Pb, which stands for "plumbum," which is the Roman name for lead and the word from which we get the word "plumber." The Romans used lead in their water and sewer piping. The symbol for gold is Au, which comes from the Latin word "aurum," and silver is Ag, for "argentum." The symbol for mercury is Hg, derived from the Greek word "hydrargyrum"; iron is Fe, for "ferrum"; tin is Sn, for "stannum"; and potassium is K, which comes from "kalium."

Still, most of the letters used for chemical symbols make sense intuitively. For example, H is for hydrogen and C is for carbon.

160. How do pay pool tables know which is the cue ball that needs to be returned?

A system of chutes connects to the table's six pockets. Gravity causes the pool balls to roll along the chutes until they reach the ball return. In a commercial, or pay-to-play, table, the balls line up in single file in a trough. You can typically see them through a piece of Plexiglas. Placing your coins in a slot trips a lever that allows the balls to roll out onto the playing surface. Let the game begin!

Let's say a player "scratches" by accidentally pocketing the cue ball. It needs to come out so the game can resume. There are two ways of getting the cue ball back in play.

The most common method is to use an oversize cue ball. It is separated from the rest of the pool balls by a radius-gauging device. The larger cue ball is directed to a separate chute, where it falls into an opening on the side of the table. I measured a cue ball to be 5.93 centimeters in diameter, whereas the rest of the pool balls had a diameter of 5.73 centimeters. So the cue ball is about two millimeters, or about one-sixteenth of an inch, larger in diameter.

Some commercial pool tables use a magnetic ball. A cue ball is made with a magnet in its core. As the magnetic ball passes by a detector, the magnet triggers a deflecting device that separates the cue ball and sends it to the opening on the side.

When pocket billiards became popular in England in the 1800s, people would place bets in a common fund, or pool, that would pay out to the winners. So a "poolroom" came to refer to a place where you could play pocket billiards.

161. How do they keep the eternal flame burning on JFK's grave?

President Kennedy was assassinated on November 22, 1963. Jackie Kennedy, widow of the slain president, decided she wanted an eternal flame on his grave that was similar to the Flame of Remembrance that burns at the Tomb of the Unknown Soldier beneath the Arc de Triomphe in Paris, France. It fell to a colonel who was an army engineer to get the flame ready within a day. His staff found a luau lamp in a local electrical shop, while his crew of engineers welded metal strips as a base and another crew laid a gas line from a propane tank to the burial site.

On the day of the funeral, November 25, 1963, Mrs. Kennedy used a candle to ignite the flame. Once, the flame went out when a visiting Catholic school group threw holy water on it. The cap came off the bottle, water poured onto the flame, and extinguished it. It was immediately relit. In August 1967, an unusual heavy rain put the flame out. A cemetery official relit the flame.

On March 14, 1967, Kennedy's remains were interred at a permanent site, which was only a few feet away from the original. At that time a new torch was designed and fed from an underground natural gas line. The new torch is in the center of a five-foot circle of granite. An electric spark will relight the torch automatically if it should ever go out.

Both an infant daughter, who was born and died on August 23, 1956, and an infant son, Patrick, born prematurely, and who lived from August 7, 1963, to August 9, 1963, were buried at the site on December 4, 1963. When Jacqueline Kennedy died of cancer in 1994, she was laid to rest next to her husband and children.

162. Why are US elections held on Tuesdays?

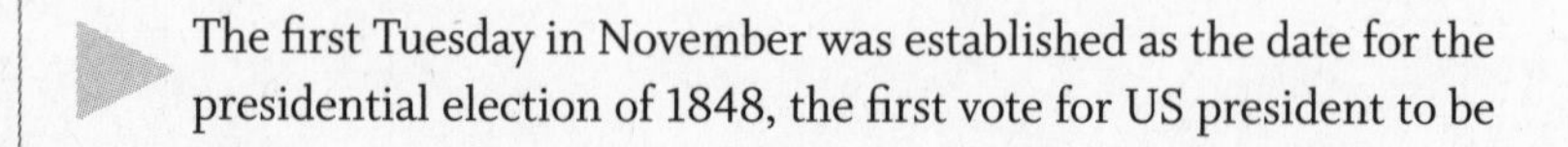

The first Tuesday in November was established as the date for the presidential election of 1848, the first vote for US president to be

held on the same day in all states. Zachary Taylor, known as Old Rough and Ready, ran as a Whig candidate. He was the last president to own slaves while in office and had a brilliant forty-year military career. Taylor died in office after serving sixteen months.

The first Tuesday in November is standard now, but it wasn't always that way. In provincial America, before the Revolutionary War, voters had to travel to the county seat to cast their ballots. Elections in northern colonies were in the spring and fall so that snow wouldn't prevent far-flung voters from arriving on time. By November, the crops were in. Roads were mostly dry and passable, which was important because it could be a trek of twenty-five miles or more on horseback. Polls would stay open for several days. Election Day was a big deal, with big crowds and much drinking and carousing. And crowd control was a problem—no rooms in the inns!

Mondays and Tuesdays were the more popular days for elections after the Revolutionary War. After 1776, more polling places were established so that would-be voters didn't have to travel long distances to the county seat. Each locality could set the date and hours of polling.

But it soon became apparent that Monday polling was not good because many people would have to start travel on Sunday. That was a day of worship, not travel. By the mid-1850s, Tuesday was the most popular day, and it soon became law to hold elections on that day.

Congress did not want elections held on November 1, because that was All Saints' Day, a holy day in the Catholic Church. Also, merchants did their bookkeeping for the previous month on the first day of the month. So the law specified "the first Tuesday after the first Monday."

November seems to be a good time of the year for national, state, and local elections. Crops are in, school is in session, summer vacations are over, Thanksgiving is a few weeks away, winter snows haven't arrived, and stores have been gearing up for Christmas for months!

There was a saying: "As Maine goes, so goes the nation." Maine held its elections for governor on the second Monday in September all the way up to 1957, and most of the time the president elected that November was of the same party. So it was thought that Maine provided a political barometer, if you will, of how the rest of the nation might vote.

There has been talk of moving Election Day to a weekend because it might be more convenient, so more people could vote. And every country has different rules. In the United Kingdom, they vote on Thursday. Germany votes on a Sunday or holiday. Voting in Australia is compulsory.

163. Why are there 5,280 feet in a mile?

We can thank merrie olde England for the statute mile of 5,280 feet. It was an act of the English Parliament in 1592–1593 (hence the name "statute") that codified the mile. For purposes of surveying, a mile is 1,760 yards, or eight furlongs. Each furlong is ten chains; each chain is four rods. The rod, then, is 5.5 yards, or 16.5 feet.

But it goes back further than Elizabethan times. The nasty Romans ruled Britain for centuries. The Romans had a measurement known as *mille passus,* meaning a thousand paces. A pace was two steps, or about five feet. Multiply five by one thousand and you get five thousand feet. When the Romans left England in 410 AD to defend the empire around Rome, the remaining Brits were in a quandary. They had their own unit, an agricultural measurement called the furlong. The furlong was how "long" a distance a horse or ox could plow a "furrow" before it needed a rest. They figured that was 660 feet. So eight furlongs were in a mile, and six hundred and sixty feet multiplied by eight gives 5,280 feet in a mile. The English had to decide on the Roman 5,000-foot mile or their own 5,280-foot mile. Property deeds at the time were in furlongs, and Queen Elizabeth I put her foot down (no pun intended) and demanded that the English Parliament make 5,280 feet equal to a mile.

These days our country is increasingly using the nautical mile at sea and in the air. It is 6,080 feet and defined as a minute of arc along the meridian of the Earth. A nautical mile per hour is known as a knot. All aircraft and ships at sea, both civilian and military, are using knots for speed and not miles per hour. There is about a 15 percent difference. For example, a speed of 115 mph is the same as 100 knots. And we're now using metrics in our track-and-field events. We have the 100-meter, 200-meter, 400-meter, 800-meter, 1,600-meter, and 3,200-meter running events. The 100-meter dash previously was the 100-yard dash; the 400-meter dash was 440 yards; the 800-meter run used to be the half mile, or 880 yards; and the 1,600 run was called the mile.

164. How do microwave ovens cook food?

Microwave ovens are actually radar sets. Radar was developed prior to World War II in both England and the United States and is credited with giving the British a fighting chance in the Battle of Britain in 1940–1941 (see page 181). Early radar workers sometimes got quite a surprise. When radar technicians moved in front of a working radar antenna, the candy bars in their pockets would melt. They also noticed tiny sparks coming from some of the metal tools.

The heart of a microwave oven is a fist-size vacuum tube called the magnetron. A magnetron is an electronic device that creates electromagnetic waves by using electricity to heat a filament wire. The result is that electrons wiggle and emit waves of about 2,450 MHz. Microwaves are exactly like visible light waves, except you can't see them. Each wave is about five inches in length, too long for the eye to detect. In the microwave oven, the beam of waves strikes a fan that distributes the waves evenly throughout the oven.

Most foods that need cooking or heating contain a lot of water. Water molecules are composed of bipolar hydrogen and oxygen atoms (see page 245). That is, the oxygen atom is slightly negative and the hydrogen atoms are slightly positive. When the water molecule is struck by microwaves, it vibrates wildly and rapidly back and forth, rotating first one way and then another. This rotation happens millions of times each second. All this twisting causes friction that heats up the food.

It's a different story for the dishes. The molecules in dishes, made of paper, glass, ceramic, or plastic, have no or very little water content, so there are practically no polar water molecules to twist or rotate and cause friction. Most of the heat that the dishes receive comes from the food that is being heated. So an empty dish in a microwave does not get very hot.

The waves from a microwave oven can penetrate to a depth of about two centimeters, or nearly one inch. So the amount of microwave radiation reaching the center of a slab of meat from all sides is more than that which is absorbed by an outside layer. This is another way of saying that microwaves cook from the inside out. The center of a steak can "get done" while the outside is still pinkish. Traditional gas or electric ovens heat by conduction, which means the outer part is cooked first and the interior is cooked last.

Why no metal in a microwave oven? Metals reflect waves, just like mirrors. Remember, microwaves are the same as radar waves used by police and troopers to catch speeders. Those radar waves bounce off my car (remember, I've gotten six speeding tickets to prove it) and return to the police radar receiver. The excess energy bleeds off and ionizes the air. That's the source of those tiny bluish lightning bolts you see coming off the errant fork! Metal is a conductor of electricity, so a metal fork will try to get rid of the built-up electric charge just like a charged-up cloud. One can see sparks flying at any sharp edge where there is a high electric-field gradient; it's a fire hazard. Some microwaves have metal trays, but they have rounded corners and no sharp edges, which prevents a build-up of voltage. Sharp edges cause discharges of electricity, similar to lightning bolts.

Sometimes we actually want metal to heat up in a microwave. For example, the sleeve around Hot Pockets has a metallic coating of aluminum foil. The microwaves cause currents to flow through the metal film, making it hot and warming the Hot Pocket. Without the metal film, the outside of the Hot Pocket would remain doughy and chewy.

165. Why do car wheels sometimes look like they're going backward?

This phenomenon is called the wagon-wheel effect, or stagecoach-wheel effect, because it is commonly seen in old oater westerns, where the stage pulls into town and it looks as though the spokes of the wheels are turning in the wrong direction. When the stagecoach slows down a bit, the wheels appear to be turning in the correct direction. You can see the same phenomenon in airplane propellers, helicopter rotors, and modern cars whose wheels have wire spokes that mimic stagecoach wheels. The effect is an optical illusion known as the stroboscopic effect. We see this illusion most commonly in movies, because of the way cameras work. Standard movie cameras take twenty-four individual pictures every second, which limits the visual information we see when we watch films.

Consider the following: A hypothetical wheel has only one spoke and is turning clockwise while a movie camera films it at the normal rate of twenty-four frames per second. The wheel is turning at a rate of twenty-four times every second, and the film camera shutter opens each time the spoke is at the twelve-o'clock position. When we viewed the film, the wheel would appear to be stationary, since we would only ever see the spoke at the twelve-o'clock position. If the rotating wheel slowed slightly, so that the spoke made it only to the eleven-o'clock position, the wheel would be behind the position it was in when the previous frame was taken. It would appear that the wheel turned slightly counterclockwise and so had rotated backward, even though, in reality, it was still turning clockwise. Let's say the wheel rotates slightly faster and would make more than one rotation before the camera shutter opened, so that the spoke would appear at the one-o'clock position the next time a frame was captured. The wheel would seem to be moving clockwise, in its forward direction, but much slower than its actual rotation rate. Of course, real wheels have many spokes. How we view the rotation of the wire spokes on a car or stagecoach wheel depends on the position of those spokes and when the camera shutter takes a picture.

I should add that this is a very simplified model; researchers are of varying opinions on what exactly is going on with our perception of the wheel. The wagon wheel phenomenon is quite complex, with terms such as "beta movement," "Schouten's theory," and "temporal aliasing theory" to describe some of the more technical aspects.

166. Which came first, the chicken or the egg?

Every generation has argued this question. It is an ancient dilemma going back to the times of Aristotle (384–322 BC) and Plutarch (about 46–120 AD). Aristotle took the easy way out, concluding that both the chicken and the egg must have always existed. Aristotle, like Plato, believed that everything on Earth first had its being in spirit.

In science and engineering, the situation is known as circular reference, in which a parameter must be known to calculate the parameter itself. In other words, one must know something to calculate that same something.

Stephen Hawking, the famous astrophysicist who is often called the successor to Albert Einstein, has argued that the egg came before the chicken. Hawking, an ardent thinker in his own right, is an adherent of Christopher Langan. Both Hawking and Langan are said to have IQs approaching 200. Langan has developed a "Cognitive-Theoretic Model of the Universe." He tackles the chicken-and-egg problem in "Which Came First?," one of the philosophical essays in his book, *The Art of Knowing*.

A literal interpretation of the Bible would put the chicken before the egg. To quote Genesis: "And God blessed them, saying, be fruitful, and multiply, and fill the water in the seas, and let fowl multiply in the earth."

Hinduism and Buddhism hold that there is a wheel of time, meaning that there is no first in eternity. Time is cyclical. There is no creation, so neither the egg nor the chicken came first.

Here is another argument: Chickens came about from non-chickens through small changes, or mutations, in the DNA. Prior to the first true chicken, there were non-chickens. The DNA changes came about in cells housed in the egg. So the egg came first.

In July 2010, British scientists, using a supercomputer, claimed to have come up with the final and definitive answer. They identified the protein, ovocleidin-17, that is required to speed up the production of eggshell within the chicken. In twenty-four hours, an egg is ready to be laid. An egg cannot be produced without the chicken. So that settles it, once and for all. The chicken came first.

That's my answer and I'm sticking to it!

167. Which will cause more damage: Running my car into a wall at 60 mph, or crashing my car, going 60 mph, into another car coming at me at 60 mph?

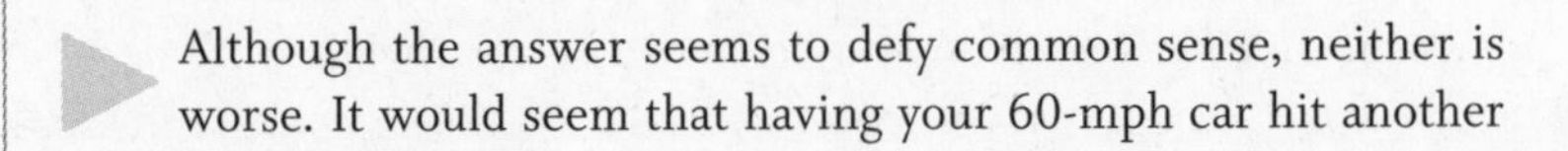

Although the answer seems to defy common sense, neither is worse. It would seem that having your 60-mph car hit another

60-mph car head-on would be worse. But think about just *your* car. It has to come to a stop by taking all of the kinetic energy (energy of motion) and turning it into the work of crushing the car, also some sound energy, and heat energy, whether by hitting a car or by hitting a wall. You can calculate the kinetic energy by the formula $KE = \frac{1}{2} mv^2$, where KE is the kinetic energy, m is the mass, and v is the velocity (speed). But the other car has to come to a halt when it hits your car, too. Yours doesn't borrow any kinetic energy from the other car, because each car is going to use all of its kinetic energy to stop itself.

A simple experiment might help illustrate the point. Suspend a ball from the ceiling by an elastic string, pull it back, and let it hit the wall. It will bounce back a certain amount, which depends on the bounciness of the ball. Now, if you let the ball hit another ball coming at your ball at the same speed that your ball is traveling, you would find that your ball bounces back the same amount as it did when it hit the wall. This assumes that the ball and wall are made of similar materials.

It may be safer to do the "two balls" experiment rather than using real cars!

168. If I dug a hole through the Earth and slid through, what would happen when I reached the other side?

If you dug a hole or tunnel all the way through the Earth, you'd have an eight-thousand-mile-long hole. You'd then have to somehow take all the air out of the hole, so that you'd encounter no air resistance, and then: You jump into the tunnel, and down you go.

You would fall and gain speed all the way down to the center of the Earth, where there is no gravity. But you're going very fast, at a speed of about 18,000 mph, and your momentum keeps you going straight out the other side, although you do lose speed all the way from the center back to the other side. It would take you about forty-five minutes for a one-way trip.

If you didn't grab the edge, you would fall back down, going faster and faster until you reached the center of the Earth. Then you would slow down again and get back to your original jumping-in point in another forty-five minutes. Total round trip would be about ninety minutes.

Now, we're assuming you had no air resistance in this tunnel and didn't get burned by the molten core of the Earth. Under these hypothetical circumstances, you would just oscillate back and forth from one end of your tunnel to the other. In this idealized scenario, your body would be following the rules of a pendulum or spring in a process called simple harmonic motion.

However, digging such a tunnel through the entire Earth is beyond the engineering capability of humankind at this point. You would have to dig through as much as thirty-plus miles of continental crust, another eighteen hundred miles of mantle, an outer core of liquid iron that is about 10,000°F, and then an inner core of solid and molten metal.

The deepest holes ever drilled on Earth include a forty-thousand-foot well on Russia's Kola Peninsula, close to the Norwegian border, and wells of similar depth in Oklahoma and the Middle Eastern country of Qatar. The Russian well is part of a scientific operation to better understand the crust of the Earth, while the Oklahoma and Qatar wells are for oil. Still, these wells, or boreholes, are about eight miles deep, and that is not anywhere near the roughly eight thousand miles required to bore all the way through the Earth.

169. Is there a particular reason why interstate speed limits are usually about 65 mph?

There are several good reasons to have a 65-mph speed limit. First, speed kills. The damage, injuries, and deaths in accidents are proportionate to the kinetic energy involved. Kinetic energy varies as the square of the speed, or velocity. (The formula for kinetic energy is $KE = \frac{1}{2} mv^2$, where m is mass and v is velocity.) This means that a 60-mph crash is not twice as bad as one at 30 mph, but rather two squared, or four times, as bad. A 90-mph crash is three squared, or nine times, as bad as a 30-mph crash. So a speed of 65 mph allows people to get places in a timely manner while protecting them from exponentially more dangerous accidents.

Higher speeds require longer stopping distances, and it is not a linear progression. In other words, going twice as fast requires more than twice the stopping distance. Total stopping distance is based on "thinking" distance and braking distance. It takes a fraction of a second to perceive a hazard and a fraction of a second to react before your foot hits the brake. That's the thinking distance. It takes time, of course, for the car to brake to a stop. Total stopping distance for a car going 30 mph is 75 feet. A car traveling at twice the speed, 60 mph, requires 236 feet to stop. That is three times the stopping distance. Consult this website for a chart of braking/stopping distances: http://www.csgnetwork.com/stopdistinfo.html.

Third, gas mileage is greater at slower speeds. That's because air resistance also varies as the square of the velocity. Gas mileage tapers off significantly at about 55 mph. The OPEC oil embargo of the mid 1970s saw gas prices shoot up and a national speed limit of 55 mph. Highway deaths decreased a little and gas mileage increased.

Interstate speed limits in most northeastern states are 65 mph. Most southern states have a 70-mph limit. Most western states have set the speed limit on the interstate system at 75 mph. Utah has an 80-mph limit, and a stretch in Texas is set at 85 mph.

170. How is paper made?

On anybody's list of top ten inventions, you will find paper and its attendant printing press. Yes, you will also find steam engine, clock, antibiotics, tools, automobile, lightbulb, laser, atomic energy, and a host of others. But it seems like paper makes it onto every list.

The inventor of paper was a Chinese court official, Cai Lun. He made paper by taking the bark from a mulberry tree, along with rags and bamboo fibers, mixing them in water, pounding the mixture, and pouring it on a flat piece of woven cloth. When the water drained, the wood fibers were left on the cloth as a thin sheet. Cai Lun presented it to the emperor Ho Ti as a substitute for silk, which was too valuable to write on. The year was 105 AD. Modern papermaking is modeled on the Cai Lun method.

The first "kind of" paper dates to around 2600 BC. The Egyptians used strips from papyrus reed, which they moistened, made into a crisscross pattern, then hammered into sheets and pressed them dry. Our word for paper comes from "papyrus."

When Johann Gutenberg invented the printing press in about 1450 AD, the demand for paper soared. From that moment on, most anybody with a little money could become literate and educated, own books, and build up a library.

These days paper is made from fast-growing trees such as fir and pine and some hardwoods. The wood is pulverized, mixed with water, heated, cleaned, whitened if meant for certain uses, and pressed into sheets. Unbleached paper is used in grocery bags and packaging.

Some say that the electronic media have reduced the need to use paper in books, newspapers, and magazines, which may be true to some extent. But *USA Today* still has a circulation of close to 2 million, and plenty of people are still buying print books. Plus, there is nothing quite like the feel of a newspaper or book. Even those electronic-ink readers like Kindle and Nook can't come close to the feel of paper.

171. Why are manhole covers round?

This subject took on a life of its own when Microsoft began asking this question in job interviews. The inquiry was meant to test the prospective employee's psychological makeup, poise, and ability to think on one's feet, rather than to elicit a correct answer.

For all manholes, there is a "lip" around the rim of the hole, holding up the cover, which means that the underlying hole is smaller than the cover. A round manhole cover can't fall through its circular opening because no matter how you position it, the cover is wider than the hole. But a square, rectangular, or oval manhole cover could fall in if it were inserted diagonally in the hole.

There are other reasons, too. Manhole covers are round because manholes are round, and manholes are round because it is the best shape to resist the compression of earth around it. Plus, a round shape is the most efficient use of materials.

Furthermore, round castings are easy to machine compared to square or rectangular castings. Manhole covers are also heavy, weighing in at over fifty pounds, and a round manhole cover can be rolled to its hole, which is easier than carrying it. Last, a worker doesn't need to line a round manhole cover up with the hole to put it back in place.

Manhole covers go back before the time of Christ. The ancient Romans used square sewer grates made of lime sandstone. Today, manhole covers are heavy enough so that cars can pass over them without any problem. There is, however, one exception: very low-to-the-ground modern race cars create enough vacuum to lift a manhole cover off the ground. It is the classic Bernoulli effect (see page 152), with fast airflow creating low pressure above the cover, so the normal pressure beneath it is high enough to lift it straight up. This scenario came true in Montreal in 1990 during the Group C World Sportscar Championship race. A Porsche lifted a manhole cover, which struck another Porsche behind it. The trailing Porsche caught on fire, and the race had to stop.

172. **Why do we have leap years?**

We have a leap year every four years because the Earth does not revolve around the Sun in a whole number (no fractions) of days. There is no reason why it should; it would be a freak accident of nature if the time it takes the Earth to rotate on its axis were an exact multiple of the time it takes to make one full orbit around the Sun.

The Earth needs 365.25 days to go around the Sun. That extra quarter, or one-fourth, of a day, added up four times, means that we need to add a day to the yearly calendar every four years. That extra day, every four years, is February 29 in the Gregorian calendar that we all follow in the Western world. We had leap years in 2008 and 2012, and the next ones will be 2016, 2020, and 2024.

However, it gets a bit more complicated. Technically, the Earth does not take 365.25 days but rather 365.2422 days for one orbit. (We could also say this as 365 days, 5 hours, 48 minutes, and 46 seconds). That extra day every four years overcompensates for the difference. Here's the fix. Over a period of four hundred years, the totaled errors amount to three extra days, so the calendar leaves out three leap days every four hundred years. There are February 29s in the three century years (integer or whole-number multiples of one hundred) that are not whole number multiples of four hundred. The year 1600 was a leap year. The years 1700, 1800, and 1900 were not leap years. However, the year 2000 was a leap year. The years 2100, 2200, and 2300 will not be leap years. The year 2400 will be a leap year. The year 2500 will not be a leap year, and so on.

There are a few simple steps to determine a leap year. First, the year's number must be evenly divisible by 4. If the year can also be evenly divided by 100, it is not a leap year, unless the year is also evenly divisible by 400, in which case it is a leap year. If this seems complicated, and it is, just get a calendar from Barnes & Noble; they are sure to have it done properly. That's what I plan to do in 2100!

The name "leap year" is derived from the fact that a fixed date on the calendar advances one day of the week from one year to the next. However, in a leap year, the day of the week will advance two days, from March forward, (no pun intended) because of the extra day of February 29. For example, Christmas Day in 2010 was on a Saturday, on a Sunday in 2011, on a Tuesday in 2012, and on a Wednesday in 2013. Christmas Day "leaped over" from Sunday to Tuesday in the leap year of 2012.

It is a tradition in Britain and Ireland that women may propose marriage on leap years. In Greece, marriage in a leap year is considered unlucky. In some countries, if a man refuses a marriage proposal from a woman on leap day, he is expected to pay a penalty, such as a gown or money. In other countries, if a man turns down a marriage proposal on leap day, he is expected to buy the woman twelve pairs of gloves, supposedly to hide the embarrassment of not wearing an engagement ring.

6

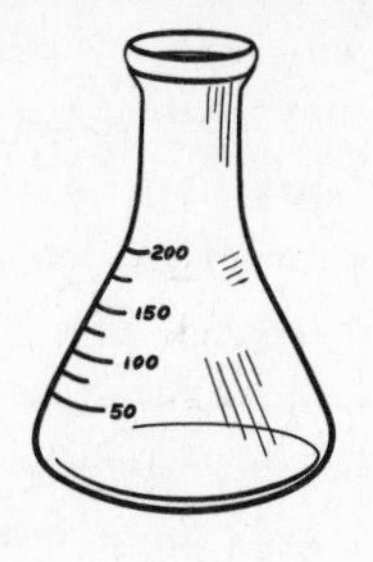

Captivating Chemistry

173. How can carbon monoxide be so dangerous but have no smell?

Carbon monoxide (CO) is produced by burning material containing carbon. Whereas most deadly gases have some kind of odor that can give us a warning of impending danger, you can't see, smell, or taste CO, but it can kill you. Known as the silent killer, CO is the leading cause of accidental poisoning deaths in America.

According to the Consumer Product Safety Commission, 170 Americans die each year from accidental exposure to CO. In addition, CO causes ten thousand hospital visits per year. Several hundred people commit suicide each year by intentional poisoning by CO. Cases of accidental or intentional CO poisoning by car exhausts are way down, however, because modern cars have catalytic converters, which eliminate 76 percent of the CO produced. Common sources of CO include furnaces, heaters, wood-burning stoves, car exhaust, camping stoves, forklifts, gas-powered generators, and welders. The usual cause of excess CO is improper ventilation.

Carbon monoxide binds to hemoglobin—the red blood pigment that normally carries oxygen to all parts of the body—over two hundred times more strongly than oxygen, so when it is inhaled, it takes the place of oxygen in hemoglobin. In reality, the carbon monoxide is causing oxygen starvation. Early symptoms include headache, nausea, and fatigue, which adds to the danger, since these symptoms are often mistaken for the flu. But then the consequences escalate; prolonged exposure can lead to brain damage and, ultimately, death.

Carbon monoxide detectors, often combined with smoke detectors into a single unit, can be bought for about twenty dollars. Many states require a detector in newly constructed houses. The most commonly used sensors employ thin wires of tin dioxide semiconductor on an insulating ceramic base. Carbon monoxide reduces the electrical resistance and allows a higher current, which is monitored by an integrated circuit and triggers an alarm.

Natural gas, widely used to heat our homes, offices, and factories, is similar to CO in that it has no smell and can kill. Gas companies add the odorant Scentinel A, which is made of ethyl mercaptan, to the natural gas so we can be warned about any leaks.

174. What is it about onions that makes us cry?

Slicing through an onion breaks open a number of cells, which releases enzymes. The escaped enzymes, allinase and lachrymatory-factor synthase, decompose some of the other released substances, called amino acid sulfoxides. This reaction forms unstable sulfenic acids, which stabilize into a volatile gas. When this gas reaches our eyes, it reacts with the water that is intended to keep our eyes moist. The sulfenic acids mix with the tears in our eyes to form sulfuric acid, which is the same toxic stuff that is in our car batteries. Nerve endings in our eyes pick up this irritant and send a message to the brain, which then passes a message to our tear ducts that says, "Dilute that irritating acid to protect my eyes." So crying or tearing is a protective measure.

The best way to deal with this phenomenon is to move away from the sliced onion instead of standing directly over it. There are also some techniques in the kitchen that you can employ. You can cook the onion first, before slicing it; the cooking process inactivates some of the enzymes. You can refrigerate it first. You can wear contact lenses or goggles. You can use a fan to blow away the fumes. You can slice the onion while running tap water over it or cut it underwater, because the water will react with some of the released gas before it reaches your eyes. Steam from a kettle will also do the trick.

When I was a kid on the farm, we pulled up horseradish from the garden every springtime. I have vivid memories of putting the horseradish into a grinder and turning the crank by hand. My, how the tears did flow! That horseradish grew wild in the garden, and it was mighty powerful stuff. We would mistakenly wipe our eyes with our hands and, of course, make it much worse.

Chalk up sulfides in onions for the "crying" mechanism. For horseradish, we can blame two active ingredients: sinigrin, a glucosinolate, and myrosinase, an enzyme. For me, onions and horseradish making us cry is a powerful reminder of how our body reacts to dangers and how our brain orchestrates chemical changes to protect us.

175. **What makes the colors in fireworks?**

There is much chemistry and physics in fireworks; the color of fireworks, for one thing, is all in the chemicals. Your basic fireworks have been around for hundreds of years. The kind we see on the Fourth of July or New Year's Eve are shot from a mortar-like steel tube. The shell's fuse burns while it rises to the proper altitude, and a time-delay fuse ignites the bursting charge, which consists of black powder (charcoal, sulfur, and potassium nitrate). Simple shells are paper tubes or spheres filled with "stars" (wads of clay dough soaked with chemicals) and the combustible black powder. The black powder ignites the stars and throws them outward in every direction in the beautiful starburst pattern that we so often see.

Every element has a distinct signature, or spectrum. That signature is determined by the electron structure surrounding the nucleus of each atom. When any element or compound is burned, atoms are given energy and the result is that electrons move to orbits farther from the nuclei. When those electrons go back to their ground, home, or steady state, that atom emits a little piece of energy called a photon. Photons are light, and the color of that light is determined by which orbits the electrons return to.

The easiest colors to get are the longer wavelengths, such as red, orange, and yellow. The longest wavelengths of red have the least energy of any of the visible wavelengths, or colors. Burning strontium chloride ($SrCl_2$) yields red, calcium chloride ($CaCl_2$) gives orange, and various sodium salts produce yellow. Barium chloride (BaCl) gives some pleasing greens, and cupric chloride (CuCl) shines into the blue region. Turquoises and ocean greens are very hard to produce. A deep blue is the most difficult color to obtain. The chemicals necessary for a deep blue are unstable and problematic when placed in fireworks.

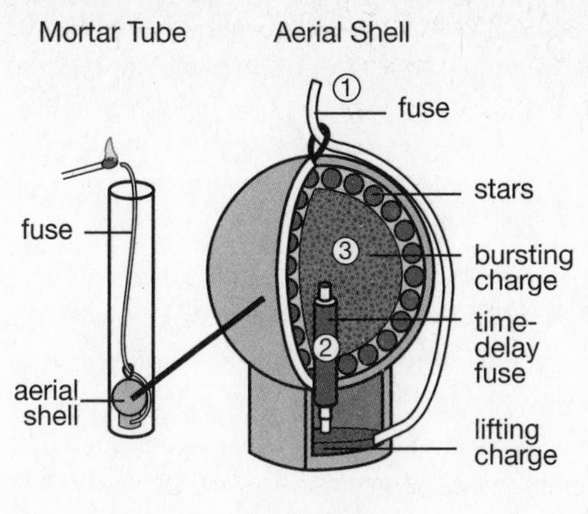

① The fuse ignites the lifting charge, which explodes and shoots the shell out of the mortar tube.

② The time-delay fuse burns slowly from the bottom while the shell gains altitude.

③ When the flame reaches the bursting charge, the shell explodes and the burning "stars" scatter.

Fireworks also contain aluminum, iron, steel, zinc, and magnesium, which create bright, shimmering sparks. The metal flakes heat up until they become incandescent and shine oh-so-brightly in the night sky. Most fireworks are actually formulations of various chemicals.

One of the biggest problems involves safety in shipping and storing. The chemicals cannot be too volatile, or they will never reach their destination. Shelf life is another headache; chemicals can decompose over time. Scientists have spent years trying to get high enough temperatures in the burst. Creating fireworks that produce brighter and deeper colors is both an art and a science. Takeo Shimizu's work *Fireworks from a Physical Standpoint* is a good read on fireworks.

176. If water is made of hydrogen and oxygen, why doesn't it burn?

Water is the "ash" left over from burning hydrogen and oxygen. When anything burns, the atoms are rearranged into a lower-energy state, because they give up some of their energy in the form of light and heat. We see that transformation into the lower-energy state as a flame or an explosion. So when the molecules have given up all their excess energy, they cannot burn. For example, wood can burn, but wood ash cannot. The ash is at the lowest possible energy state.

But there is more to the story. Wood, as well as fossil fuels, such as coal, gas, oil, and the oil contained in shale, contains a lot of carbon. Burning them causes the carbon to combine with oxygen in the air to make water vapor and carbon dioxide, neither of which can burn. Carbon dioxide is a greenhouse gas that traps heat in the atmosphere, which leads to global warming.

If these fuels burn without enough oxygen, some of the carbon forms carbon monoxide. Carbon monoxide is a very poisonous gas that burns with a blue flame (see page 232). Carbon monoxide, deadly for us to breathe, is a major industrial gas. It is used to produce detergents and acetic acid, which goes into vinegar and changes coal to gas, which provides us with a valuable fuel source.

Hydrogen, when it "burns" with oxygen, forms water. Hydrogen is the cleanest-burning fuel known. The best energy cycle possible would be to use solar energy to break down water into hydrogen and oxygen, a process called electrolysis, then use the hydrogen as a fuel.

177. How is glass made?

Glass is a hard, brittle, and transparent solid. About 90 percent of the glass we use is soda-lime-silica glass. It is 75 percent silica, or just plain sand. The "soda" refers to sodium carbonate, and the "lime" comes from limestone, which is mostly calcium carbonate. The ingredients are put into a gas-fired furnace. Window glass is made by having the molten, liquid glass flow over a molten tin bath in a continuous sheet. Nitrogen gas pushes on the top surface to make it smooth.

Glass melts, or turns to a liquid, at a temperature between 1,400°C and 1,600°C, or about 2,500–3,000°F. The melting temperature depends on the composition of the glass. Glass does not melt at an exact temperature. Unlike ice, which melts at a precise temperature of 32°F, or 0°C, glass goes through a phase-change transition. Glass gets softer and softer as its temperature increases, until it can start to flow. That's a nice property of glass, because glass can then be molded into any shape we desire.

Ingredients are added to the soda and lime to change the properties of the glass. Lead or flint causes glass to sparkle. Boron is added to glass to change its thermal properties; this makes borosilicate glass suitable for cooking. We also use a lot of borosilicate glass in the science lab. It has a very low thermal expansion rate and does not break when a very hot or very cold liquid is put in it. Pyrex is a brand name for the same kind of glass. Lanthanum oxide has light reflective properties and is used in eyeglasses. Iron is put in window glass to absorb infrared energy. Take a look at the edge of ordinary window glass, and you can see the green tint. That's from the iron.

Glass used in optical equipment is generally made from low-dispersion crown glass or denser, high-dispersion flint glass. The glass is selected based on how light is bent, or refracted, through it. Dispersion is a term that describes how much light is bent or refracted as it passes through the glass.

Tempered glass is stronger than regular glass, but if it does break, it shatters into small pieces rather than the big sharp shards of ordinary glass that can cause major bodily harm. Regular annealed glass is placed on a table, which is rolled through a furnace; then the surfaces are quenched or quickly cooled with air, while the interior of the glass remains liquid, or molten. This process sets up stresses in the glass. Tempered glass is used in side and rear windows of cars and in glass doors that are not in a frame.

There is an old myth that glass flows over time. The distortions found in glass that is several hundred years old, such as the glass in Thomas Jefferson's Monticello home, gave rise to this myth. Glass is an amorphous solid that does not flow. The variations in thickness that cause distortions were due to very imperfect manufacturing techniques.

The first glass was made by nature, not man. Ancient lightning strikes hitting common silica sand instantly heated the sand, melting it and fusing the grains into glass. The result, termed a "fulgurite," takes the form of a natural glass tube. The longest fulgurite, found in Florida, is over sixteen feet long.

A black glass from volcanoes is called obsidian. It is produced when lava cools with minimal crystal growth. Chipped obsidian edges can be as thin as a few molecules, making obsidian excellent for ancient hunters' arrowheads and modern medical scalpel blades used in heart surgery. An obsidian blade can be many times sharper than a metal blade.

Tomah has its own connection to glass. Cardinal IG (insulating glass) and Cardinal TG (tempered glass) are located in the industrial park on the eastern edge of Tomah. The IG plant fabricates low-emissivity double-pane glass for windows, used mainly in homes and office buildings. A low-emissivity, microscopic, thin metal coating, applied to one of the inside surfaces of a double-pane window, controls heat transfer. It keeps the heat out while letting light in.

178. How do they know how many calories are in food?

A calorie is a unit of heat energy. Specifically, it is the amount of heat required to raise the temperature of one gram of water 1°C, or about 1.8°F. A gram is about what a raisin weighs. One calorie is equal to 4.2 joules, a common unit of energy in the sciences.

The corresponding English unit is the Btu, or British thermal unit. But even the Brits are now using calories, and Btus are going out of style. Still, in the United States, we continue to buy air conditioners and furnaces rated in Btus.

When we think of calories, we think food calories. If a doughnut has three hundred calories, that is really three hundred kilocalories (one thousand calories = one kilocalorie). Sometimes this kind of calorie is spelled with a capital C. The same applies to exercise. If your treadmill says you burned five hundred calories, that's actually five hundred kilocalories.

I like to eat old-fashioned Quaker Oats every other day for breakfast. It says on the round, silo-shaped box that one serving (a paltry half cup) has one hundred and fifty calories. If we put that forty-gram serving of oatmeal in a bowl and set it on fire, the heat would raise the temperature of one hundred and fifty kilograms of water 1°C.

Scientists measure food calories in something called a bomb calorimeter. A small cup contains the food sample, oxygen, water, a stirrer, a thermometer, an insulating container, and an ignition circuit. They ignite the calorimeter with electricity. The burning food heats up water in a jacket that surrounds the combustion chamber. By knowing how much water it is heating and how much the temperature changes, they can calculate the calorie content.

The amount of calories in food measures how much potential energy is in the food. We humans need energy to survive—to move, breathe, and pump blood. Foods are a combination of carbohydrates, proteins, and fat. One gram of carbohydrates yields four calories. A gram of protein is also equivalent to four calories. One gram of fat is worth nine calories. Our bodies "burn" those calories by having enzymes break down the carbohydrates into glucose, the fats into glycerol and fatty acids, and the proteins into amino acids. All that stuff goes into our bloodstream and is carried to our cells.

My wife does a very good job of buying healthy food. I love those Pop-Tarts, but they're pretty much banned in our house. There are about 200 calories in one Pop-Tart. But fifty of those calories are fat. In addition, hydrogen is added to the shortening (hydrogenated), and that stuff clogs the arteries.

179. How is gasoline made and how does it power cars?

Gasoline comes from crude oil, or petroleum, that thick black stuff we see in movies that comes gushing out of the ground. We call crude oil as a fossil fuel, because it comes from decaying prehistoric plant and animal life in what was once a seabed.

Crude oil consists of hydrocarbon molecules, which are long chains of hydrogen and carbon atoms. But crude oil itself can't be used for much of anything. Crude oil contains many different types of hydrocarbons, with various chain lengths, so it has to go to a refinery. These hydrocarbons have different boiling temperatures, so they can be separated by a process called fractional distillation. This involves heating up the crude oil until it turns to vapor. Then it condenses into separate fuels, such as gasoline, kerosene, lube oil, jet fuel, fuel oil, tar, wax, and asphalt.

Burning gasoline in an engine, in the presence of oxygen, makes incredible things happen. You get carbon dioxide from the carbon atoms and water from the hydrogen atoms. But you also get a tremendous amount of heat. A car engine is really a heat engine. That heat can drive pistons, which turn a shaft, which turns wheels and takes us where we want to go. And gas is the key! The best way to create motion from gasoline is to burn the gasoline inside an engine. The burning, or combustion, takes place internally, so a car engine is an internal combustion engine.

While gasoline is an amazing fuel, it is not without its problems. Car engines are not perfect; the ideal engine would produce only heat, water vapor, and carbon dioxide. But engines produce carbon monoxide, which is a poisonous gas, and nitrogen oxides, which cause smog. Engines also put out unburned hydrocarbons; these contribute to ground-level ozone, which is a health risk.

Carbon dioxide is a greenhouse gas, and, unfortunately, burning a gallon of gasoline releases twenty pounds of carbon into the air. We don't notice the carbon dioxide because it comes out of the tailpipe as an invisible gas, and too much carbon dioxide leads to global warming and dramatic climate changes.

Almost all cars and trucks use a four-stroke combustion cycle to convert gasoline into motion. The cycle of intake, compression, power, and exhaust strokes means that the crankshaft gets one power, or push, stroke for every two revolutions. The four-stroke approach is named the Otto cycle, for Nikolaus Otto, who built the first engine based on it in our centennial year of 1876.

The basic principle behind any internal combustion engine is the same as that of the potato cannon we used in physics class. Igniting a tiny amount of high-energy fuel in a small, enclosed space releases an incredible amount of energy in the form of expanding gases. In the engine, instead of throwing a potato several hundred feet, the expanding gases push a piston that turns a crankshaft that transmits the motion to the wheels of the car. Several hundred such explosions in a minute produce a smooth push on the wheels of our car.

Gasoline does contain a lot of energy. A gallon of gasoline will run a fifteen-hundred-watt space heater for twenty-four hours or your furnace for twelve hours. A gallon of gas contains the equivalent of about thirty thousand food calories, the amount in over one hundred McDonald's hamburgers. We use around 130 billion gallons of gasoline a year in the United States.

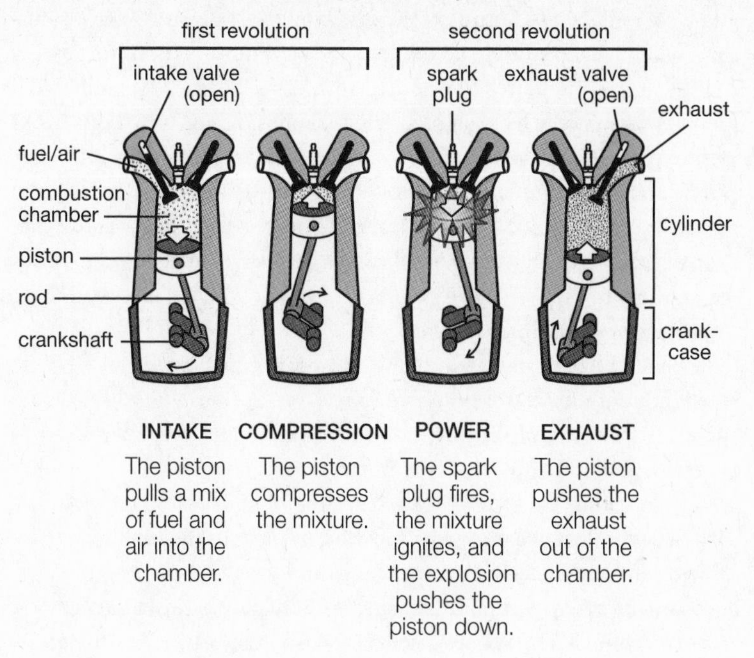

180. What is fire?

The ancient Greeks thought the world was composed of four elements: air, earth, fire, and water. Air, earth, and water are all forms of matter, but fire is a chemical process by which matter changes form. Fire is a reaction between oxygen and fuel, such as wood or gasoline. Fortunately, wood and gasoline don't just start on fire in the presence of oxygen. To get fire, the fuel must reach a certain temperature, called the ignition temperature.

Interesting things happen when wood burns. First, the temperature of the wood must get up to about 500°F. A number of things can provide this high ignition temperature; lightning, focused light, friction, or something already burning will suffice. The intense heat decomposes the cellulose material, releasing gases that we call smoke. Smoke is made up of hydrogen, oxygen, and carbon. The stuff that is left has two components: Char is almost pure carbon. Ash is everything in the wood that didn't get burned mostly calcium, iron, and potash. When we buy charcoal for that outdoor cookout, we are buying wood that has already been heated sufficiently to remove all the smoky gases. So a charcoal fire burns with very little smoke. The carbon in the charcoal combines with oxygen. This is a very slow reaction, so charcoal will burn for several hours. Both chemical reactions—wood releasing gases (smoke) and carbon producing char—produce a lot of heat, which sustains the fire. Burning gasoline, on the other hand, is a one-step process. Heat causes the gasoline to vaporize. The gasoline burns as a hot, volatile gas. There is no char and very little smoke.

To put out a fire, one must remove the heat or the fuel source or both. Firefighters dump water on a fire to remove the heat. Firebreaks and backfires are ways of removing the fire source and are often used to control huge forest fires in our western states. A backfire is a fire deliberately set along the inner edge of fire line to consume the fuel in the path of a wildfire to change the direction or force of the fire.

Fire can be a killer. It destroys homes and forests and kills more people than any other natural force, such as hurricanes, tornadoes, floods, and explosions. Fire has been used as a weapon in war for centuries.

Handled correctly, fire is also an obedient servant for us humans. It keeps us warm, cooks our food, gives us light, allows us to make metal tools, and drives power plants. Fire, by heating water and providing steam, opened the Industrial Revolution. It has made our life more comfortable and has given us an ample food supply. Fire renews our forests, by burning old forest material and breaking open the seeds for new plant life. Fire powers our transportation vehicles. The human race has tamed the flame and put it to good use.

181. What part of a flame is the hottest?

Short answer: The tip of the blue part of the flame is the hottest. But the analysis of a flame can be quite complex. Consider a cigarette lighter flame, or a candle flame, or a flame from one of those handheld butane candle lighters or charcoal grill igniters. The flame looks like a solid conical mass, but it is actually hollow. The visible outer layer is roughly the thickness of a dime, or about one millimeter. The hollow core of the flame contains the fuel gas and air pushing upward into the burning, or combustion, zone.

Here's where it gets a bit complicated. Atoms are raised to a higher-energy state during combustion as energy from the heat boosts their electrons into orbits farther from the nucleus. When the electrons drop back into their original orbits, the atoms emit light. Flame colors come from the energy released by the electrons of the atoms of burning gases; each color corresponds to a particular frequency, which is a function of the amount of energy released. Red, orange, or yellow flame corresponds to low frequency, low energy, and low temperature. Blue or violet flame means high frequency, high energy, and high temperature. So the blue part of the flame is the hottest.

The color of a hydrocarbon flame, such as that from a candle or gas, also depends on the oxygen or air that is mixed with the fuel. If the mixture is a bit lean (in fuel, so it has more oxygen than really needed), the flame will skew bluish. If the mixture is too rich (not enough oxygen), the color trends toward orange/yellow. Carbon particles form and are heated in the flame. Soot particles, made up of burned carbon, often give a black color to the outer edge of the yellow flame.

Those blacksmiths of old understood the basic concepts behind temperature and color. They knew that the hottest flame was blue-white. They also knew that they could obtain the hottest flame by making their mixture slightly lean, i.e., having slightly more oxygen than needed. That's why they used a bellows to shoot extra air into the furnace. They needed a really hot flame to melt or soften the iron so it would be malleable, which allowed them to hammer the metal into shapes.

Science labs are often equipped with Bunsen burners, which use natural gas (methane) or propane or butane (liquefied petroleum gas) to produce a single open flame. These burners get their name from Robert Bunsen, who, in 1855 developed a new and safe method of mixing gas and air. Bunsen worked with Peter Desaga at the University of Heidelberg, who had perfected a design by Michael Faraday. Students today are taught to adjust the air flow so that the flame burns blue, not yellow.

The hot gases in a flame are much less dense than the surrounding air, which is why here on Earth, flames spread upward in our gravity and are pointed at the top. In contrast, on the International Space Station, a weightless, or free fall, environment, a flame forms a beautiful ball, or sphere.

182. What is lead made out of?

Well, lead is made out of lead, an element all by itself. Lead is a bluish-white lustrous metal, very soft, and highly malleable. "Malleable" means that you can easily pound or roll it into a thin sheet. Lead is one of the few metals that are not good conductors of electricity. Lead is heavy. It is 11.3 times as dense as water. A five-quart ice cream pail of lead weighs about one hundred and twenty pounds. Lead is the end result of the natural radioactive decay of Uranium-235 and Uranium 238.

The symbol for lead is Pb, derived from the Latin "plumbum," meaning "liquid silver." It has atomic number 82 and an atomic weight of 207 on the periodic table. Lead goes way back; even the Bible mentions it, in the book of Exodus. For centuries, alchemists tried to turn lead into gold. They were never successful.

Lead is very resistant to corrosion, so it was used for centuries in plumbing. Archeologists have found lead pipes bearing the insignia of Roman emperors, and lead pipes still drain our baths today in many homes.

Alloys using lead include solder and pewter. Tetraethyl lead, used in gasoline for many years to prevent engine knock, has largely been phased out because of its toxicity. Lead is a major part of the storage batteries in cars and trucks.

Lead is excellent as a sound absorber. It also blocks radiation very effectively and makes a wonderful shield around X-ray equipment and nuclear devices (see page 275). It's used in the screens of TV sets and computer screens to reduce radiation, it helps in correcting defects in eyeglass lenses, and it's also a component in fine crystal glass. A lot of ammunition contains lead. Lead is used in sinkers on fishing lines and plumb bobs in the construction industry.

But lead is poisonous, and toxic even in small amounts. Lead poisoning can cause damage to the brain and nervous system. In adults, excessive lead causes high blood pressure, nerve disorders, kidney problems, and problems having children. Children are most vulnerable to lead poisoning, because their bodies and nervous systems are still developing. Their tendency to put things in their mouths once put them at great risk of poisoning from lead in paint chips, and even today, lead can filter out after it has been painted over. For them, lead exposure can bring on loss of intelligence, learning disabilities, and behavior problems. Lead may have many uses, benefits, and advantages in products that we use, but lead has to be monitored closely because lead poisoning is wicked stuff!

183. What makes ice float?

The simple answer is that ice is less dense than water. The question then becomes: Why is ice, which is water in solid form, lighter than water in its liquid form? Something must be happening to water when it freezes.

One molecule of water has two hydrogen atoms and one oxygen atom, or H_2O. The hydrogen atom is slightly positive, and the oxygen atom is slightly negative. At room temperature, these water molecules are loosely linked and free to move around easily, which is what defines the liquid state of matter. Once the temperature hits 32°F, or 0°C, the molecules line up in a rigid six-sided honeycomb structure, with water molecules linking up with each other in a process called hydrogen bonding (the more positively charged hydrogens attract the more negatively charged oxygens). This causes the molecules to move more slowly and take up more space. You have ice. Because the same weight or mass now takes up more space, it is less dense and floats on water.

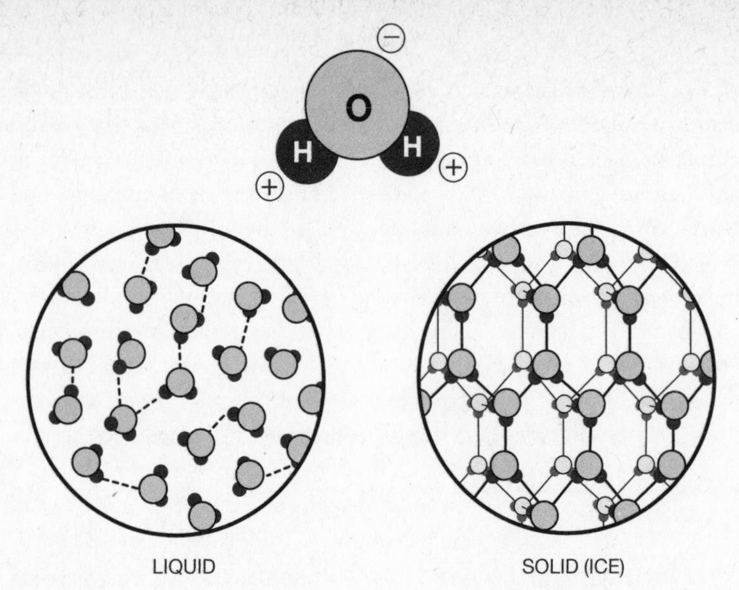

The loosely bonded water molecules are in motion and closer together.

The tightly bonded water molecules are stationary and farther apart.

We can put salt on ice to melt it. Salt is sodium chloride, or NaCl. (Technically, most salt for sidewalks is potassium chloride, or KCl, but the chemical process works the same way.) The sodium atom in a salt molecule has a slight positive charge, and the chloride atom has a tiny negative charge. A water molecule has one atom of oxygen bound to two atoms of hydrogen. The hydrogen atoms are "attached" to one side of the oxygen atom, resulting in a water molecule having a positive charge on the side where the hydrogen atoms are and a negative charge on the other side, where the oxygen atom is.

The sodium chloride ions pull and tug on the water's two hydrogen atoms (unlike charges attract) and tear the bonds apart. The ice no longer has a honeycomb structure. It has melted. In order for the melted ice to freeze again, it has to go much lower than 32°F; the salt has lowered the freezing point of the water.

Most of Earth's ice is packed into huge sheets called glaciers. The largest glacier covers most of the Antarctic continent and holds about 90 percent of the world's ice. If all of it were to melt, the oceans would rise 187 feet.

You might want try this science experiment. Fill a drinking glass with water, and then put an ice cube in the water. You can see that the ice cube floats. Now, using isopropyl rubbing alcohol that you can buy in any drugstore, fill a glass with alcohol and carefully drop in an ice cube. What happens to the ice cube? Alcohol is less dense than water, less even than frozen water, so, since the ice cube is denser than the alcohol in the glass, it sinks.

184. How do they separate oxygen from air to make oxygen tanks?

There are two major methods of producing oxygen and putting it in a bottle or tank. The most common is fractional distillation. That is a fancy way of saying that technicians turn air, which is composed mostly of oxygen and nitrogen, into a liquid by compressing and cooling it, then separate the oxygen and nitrogen by letting the nitrogen turn into a vapor, or gas, while the oxygen is left behind.

The second major method of producing oxygen is called pressure swing adsorption. This works by passing clean, dry air between two layers of zeolite sieves, which adsorb, or attract to their surfaces, the nitrogen but not the oxygen. Zeolites are tiny aluminum silicate minerals.

There is a third, but minor, way of getting oxygen. It is called electrolysis and uses direct current electricity to break down a water molecule into its hydrogen and oxygen parts. This is often done as a demonstration in a physics or chemistry class.

Oxygen is the third most abundant element in the universe, after hydrogen and helium. It is contained in water, in air, and in the crust of the Earth. At ordinary temperatures, oxygen is a free gas that has no odor, no taste, and no color. All animals and most plants need free oxygen to live. Without it, they would suffocate. By weight, oxygen makes up about 21 percent of the air we breathe. Oxygen combines with hydrogen to make water, or H_2O. And oxygen is one of the most abundant elements that constitute soil and sand. Fifty percent of soil is composed of pore space. That pore space is filled with water and air. Oxygen is in both the water and the air. There is actually more oxygen in soil than in the air we breathe!

The chemical symbol for oxygen is O, and it has atomic number 8, meaning its nucleus contains eight protons. A molecule of ordinary oxygen contains two atoms of oxygen, and its symbol is written as O_2. Another form of oxygen, ozone (O_3), helps protect us from ultraviolet radiation. That beneficial ozone layer is high in the atmosphere. But if ozone is down at ground level, it is a pollutant and a component of smog.

Oxygen is the key element in the process of breathing. Animals take in oxygen and breathe out carbon dioxide. Plants, in turn, absorb carbon dioxide and release free oxygen during the process of photosynthesis.

Oxygen is most often transported in liquid form in bulk insulated tanker trucks. You see those bulk semis go up and down the interstate system. These tankers refill bulk liquid oxygen tanks at hospitals and factories. Transporting oxygen in liquid form is much more economical than carrying it in gaseous form. Oxygen is also shipped and stored as a gas in green cylinders. You may have seen people with respiratory problems in public carrying or carting small bottles of oxygen. Oxygen bottled in gas form, along with acetylene, is used in oxy-fuel cutting torches.

When it was operating, the space shuttle used a big liquid oxygen tank and a huge liquid hydrogen tank stacked up in the external tank to feed fuel to the three main engines during launch. Then it would separate from the craft and fall into the ocean. It was a liquid oxygen tank that exploded on the *Apollo 13* mission to the Moon. The crew had to abort the Moon landing and barely made it back safely to Earth.

185. **How does yeast rise in an oven?**

Yeast is a single-cell fungus (*Saccharomyces cerevisiae*). It is the same fungus used in the fermenting of beer and wine. You can buy fresh or live yeast, which has a very short shelf life and needs to be refrigerated. You can also purchase dried versions, which you can store for long periods of time but will need to activate by heat or warm water before using it.

The ingredients in bread are flour, salt, sugar, and yeast. Yeast devours sugar and then releases carbon dioxide bubbles and a tad of ethyl alcohol.

The released carbon dioxide gas gets trapped in the dough and causes the bread to rise. This process takes several hours. The high temperature of the oven also causes the yeast to work overtime. That's why the bread dough may continue to rise in the first few minutes in the oven. The high heat kills the yeast cells toward the end of the baking process. Salt can halt the rising action. So recipes usually call for a little salt, and a little sugar to balance it. Shortening and animal fats also slow down the yeast action, so salted or buttered breads do not rise as much as white bread.

The expressions "leaven" and "leavening agent" mean any substance that causes a foaming action that lightens and softens the product. That foaming action forms pockets of gas trapped in a solid or a liquid. One of the first known leavening agents was sourdough, and its history extends back twenty-five hundred years.

Sourdough contains a *Lactobacillus* bacterial culture along with some *Saccharomyces*. Sourdough has lactic acid in it, which gives bread a tangy, or sour, taste. Sourdough bread is made by using a small amount of starter dough, which contains the culture and yeast, and mixing it with new flour and water. Part of the resulting dough is saved as the starter dough for the next batch.

Sourdough was the main bread made during the California gold rush of the late 1840s and early 1850s and the Alaska gold rush of the late 1890s. The bread was so common and so vital that sourdough became a general nickname for the gold prospectors themselves.

186. I saw about one hundred train cars go by heading west that were labeled "Molten Sulfur." What is molten sulfur and what is it used for?

Molten sulfur becomes a red liquid at a temperature of about 235°F. For shipping, sulfur is kept at 290°F to prevent it from solidifying. Sulfur is easier to handle and transport in liquid form than as a solid.

Sulfur is an important industrial raw material. It is used in the manufacture of sulfuric acid, the liquid electrolyte in our car batteries. Sulfur makes hydrogen sulfide, which is used in many types of polymers and also for removal of hair (depilation) from animal hides.

Molten sulfur is used in the making of sulfur dioxide and sodium sulfites that are used to bleach straw and wood fibers. It is also used to remove lignin, a polymer that gives structure to wood, from wood pulp for the paper industry (see page 227). Molten sulfur is also a key ingredient in tire making. Sulfur is added to rubber at a high temperature in a process called vulcanization. This is a chemical process in which polymer molecules are cross-linked to other polymer molecules to form bonds that make the rubber harder, smoother, more durable, and more resistant to chemical attacks. The vulcanization process is named after Vulcan, the Roman god of fire. Water hoses, hockey pucks, and shoe soles are a few other products to which sulfur is added. Molten sulfur is utilized in the making of black gunpowder, insecticides, and pharmaceuticals.

Hydrogen sulfide gas is often referred to as having a "rotten egg" smell. Sulfur-reducing bacteria, using sulfur as an energy source, can produce large quantities of hydrogen sulfide. These bacteria do not need oxygen, so they thrive in deep wells, water softeners, water heaters, and plumbing systems. Surprisingly, they live on the hot side of any water system. These hydrogen-sulfide-producing bacteria are responsible for bad-smelling water.

Sulfur is one of those English words that have two acceptable spellings. You will see both "sulfur" with the "f" and "sulphur" with the "ph," although the "f" spelling is more common in American English and the "ph" spelling in British English.

187. Why does a helium balloon float, while an air-filled balloon does not?

A helium balloon floats upward in air for the same reason that some things float in water: the helium balloon is less dense than the air around it, just as a floating object is less dense than the water around it. Put another way, the combined weight of the balloon and the helium inside is less than that of the air pushed out of the way (displaced) by the inflated balloon. The air in an air-filled balloon has the same density as the air around it, and the balloon itself is denser, so the balloon falls. The same rules of buoyancy, based on the Archimedes Principle (see page 179), apply to objects in the air as to objects in the water.

If you let go of a helium-filled balloon, it rises until it pops. The air pressure on the outside of the balloon lessens with altitude, the balloon expands beyond its limit and finally breaks. The helium escapes, goes into space, and is lost forever. There is very little helium in our atmosphere at any given time. So where does helium come from?

Helium is born deep inside our Earth and derives from alpha particles created by radioactive decay of uranium. These particles have two protons and two electrons, the same as a helium nucleus, and become helium atoms when they attract two electrons to orbit them. Helium is abundant in places where there is a lot of uranium ore and natural gas. Those pockets of natural gas act as sealed containers that hold the helium. The helium is extracted from the natural gas by fractional distillation and purified with activated charcoal, then cryogenically liquefied and put into cylinders for sale to the public (see page 247).

Hydrogen makes a better lifter than helium, by about 7 percent. And hydrogen is cheaper, as it comes from breaking down water with electricity. But hydrogen has the nasty habit of exploding. Any little spark will set it off. When the Nazis consolidated their power in Germany in 1933, the United States refused to sell helium to the new regime. So the Germans turned to hydrogen. The German passenger airship *Hindenburg*, which exploded at Lakehurst, New Jersey, on May 6, 1937, contained about seven million cubic feet of hydrogen.

Helium is a very small atom, and it leaks out of a balloon through the pores in the rubber, or latex. Most helium-filled balloons lose their life after a day or two. "Helium-grade" balloons are a little thicker and last longer. The best helium-filled balloons are made of Mylar and will hold helium for a week or more.

How big a balloon is needed to lift a person? This becomes a neat mathematical problem, involving the lifting force of helium, volume, and weight. Five balloons, each with a diameter of about ten feet, will lift a 150-pound person.

In 1982, Larry Walters, a thirty-three-year-old truck driver from San Pedro, California, attached forty-five four-foot-diameter weather balloons to his lawn chair and was lofted up to between fifteen and sixteen thousand feet. He intended to go up only about thirty feet, or so he said. He finally stopped his ascent after shooting out some of the balloons with a BB gun.

Walters had dreamed of flying, but the US Air Force had turned him down because of poor eyesight.

188. Both oil and water are liquids, so why don't they mix?

There are two parts to this question. First question: Why don't water and oil mix? Second question: Why does oil stay on top of water?

One liquid will dissolve or mix into another if the molecules of the two liquids have similar electric dipole moments, or strengths. A magnet, for example, has a north and a south pole, with opposite magnetic charges. Two of the same poles, such as a north pole and a north pole, repel. Opposite poles, north and south, attract. Similarly, molecules have electric dipoles, with a positive end and a negative end based on the electric charge each end carries. The charge can be greater in magnitude on one end than on the opposite end. Materials with dipoles of similar strengths dissolve in each other more easily than those with dipoles of different strengths. Water and oil have very different dipole charges, so they don't stay mixed. Liquids that do not mix are said to be immiscible.

Now to the second part of the question. Oil is less dense than water, so it stays on top of, or floats on, water. Almost all liquids made from crude oil float on water, including gasoline, kerosene, turpentine, engine oil, baby oil, mineral oil, diesel fuel, and jet fuel. All oils made from plants, such as linseed oil, castor oil, sunflower oil, coconut oil, and olive oil, are also less dense than water and, therefore, floats on top of water, too. A classic example of this phenomenon is oil-and-vinegar salad dressing; the oil and vinegar never actually mix.

Of course, crude oil itself also floats on water, and it's a good thing, too, as it makes oil spills, like the 1989 *Exxon Valdez* disaster in Alaska, a little easier to clean up. In ancient times, sailors often poured oil on ocean water to calm wave action. Such a practice would not be allowed today.

Lava lamps are made by putting those colored floating blobs of aniline, an organic compound, into another substance, usually mineral oil.

The two liquids have densities that are very close. A lightbulb in the base heats the aniline, which makes it less dense, so it floats up to the top of the lamp. Having risen farther from the heat source, the aniline cools off, becomes denser, and sinks back down to the bulb, and the whole process starts over again.

189. What causes paper to yellow over time?

Newsprint paper contains a lot of acid and lignin. Acid causes the paper to disintegrate, and lignin turns paper yellow or brown. Lignin is a polymer that binds the wood fibers (cellulose) together. To make a fine white paper, the paper mills put the wood through a chemical solvent process, which dissolves the lignin (see page 250). Pure cellulose is white, and paper made from cellulose will be white and resist yellowing.

The newspaper people don't care if their newsprint turns yellow with time. They care about cheap paper, made cheaper by not having to chemically treat it. Newsprint paper is designed to be used for just a few days and discarded.

Paper turns yellow because the lignin molecules become less stable in the presence of oxygen. The lignin absorbs more light, hence the darker color. If newsprint were kept completely away from light and air, it would stay white. But then I guess it might be hard to read it!

Sometimes we want lignin in paper. Brown grocery bags and cardboard come to mind. Lignin makes them strong and study. Not much printing is done on them anyway.

People will cut a clipping out of the newspaper and put it in their scrapbook. It might have been an obituary, an award earned, or an event they attended. After several years, they notice the newspaper article has badly deteriorated. The best thing to do is make a copy of the newspaper article using acid-free, lignin-free white paper. Even regular typing paper does a good job. It's better than pasting or gluing newsprint in albums and scrapbooks.

What about the term "yellow journalism"? The Yellow Kid was the main character in comic strips under various titles created by a cartoonist by the name of Richard F. Outcault. The comic strips appeared in Joseph Pulitzer's *New York World* newspaper and later in William Randolph Hearst's *New York Journal-American*. In a circulation war in the 1890s, both newspapers engaged in scandalous, sensational, gaudy, and cheap reporting. The expression "yellow journalism" derives from that 1890s cartoon.

190. How are the elements organized in the Periodic Table of Elements?

First, a word about elements. An element is a substance that cannot be broken down into simpler substances. An atom is the smallest particle of an element. So an element contains only one type of atom. Elements are arranged in the periodic table in order of increasing atomic number. The atomic number is the number of positive protons in the nucleus. The nucleus has two kinds of particles: protons and neutrons. Protons have a positive charge, and neutrons have no charge.

Think of an atom as a tiny solar system. The Sun is the center of our solar system, and the planets go around the Sun in near-circular paths, or orbits. The center of an atom is the nucleus, and electrons, which have a negative charge, move in circular paths—the shells mentioned before—around the nucleus.

The periodic table of chemical elements is arranged like a big grid, similar to a stack of Lego blocks. There are rows, left to right. There are columns, up and down. Each element's position on the grid depends on the way it looks and acts and primarily how much it weighs. Each of the rows is considered to be a different period. The weeks on a calendar are laid out much the same way. All the elements in each of these periods have something in common. That something in common is the number of orbital shells that hold the electrons zipping around the central core of the atom, the nucleus. Think of these shells as being like nested eggshells. These shells are given the letters: k, l, m, n, o, p, and q. Each shell can hold only a certain number of electrons. The first shell can hold only two electrons, the second can hold eight electrons, the third can hold eighteen, and so on. The maximum number of electrons in any shell is

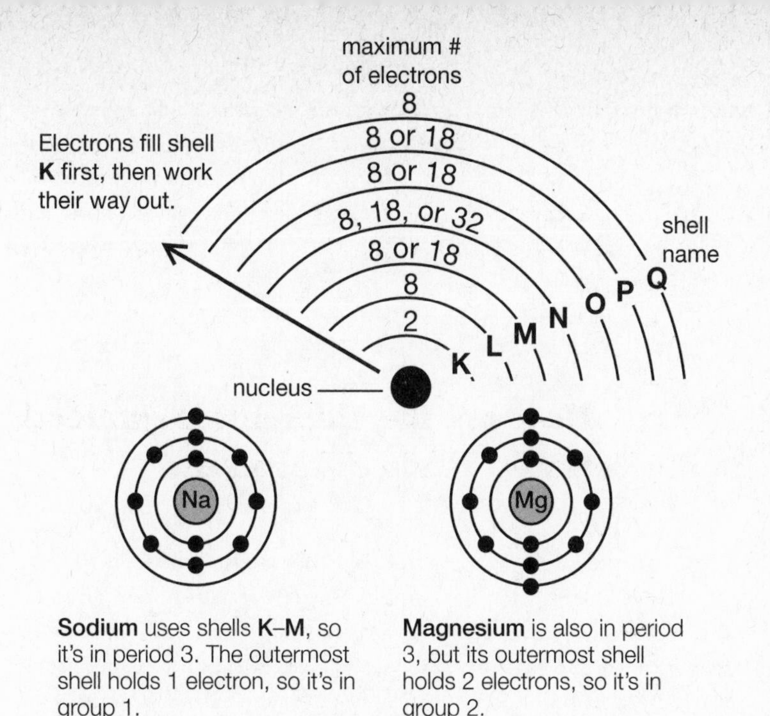

Sodium uses shells **K–M**, so it's in period 3. The outermost shell holds 1 electron, so it's in group 1.

Magnesium is also in period 3, but its outermost shell holds 2 electrons, so it's in group 2.

thirty-two.

So much for the rows...what about the columns that go top to bottom? They have a special name also. The columns are called groups. The elements in each group all have the same number of electrons in their outermost full orbital shells. Each element in the first column (group one) has one electron in its outermost full shell. Every element in the second column, or group two, has two electrons in its outermost full shell. As you move from left to right, you can determine how many electrons are in the outermost full shell. There are a few exceptions to the shell rule, like when you get to the transition elements. The transition metals are found in the middle section of the periodic table. The rules for transition metals get a little tricky because more than one outer orbit can have less than the maximum number of electrons. The number of electrons in the outer shell of an atom is important. In most elements this shell is not full, and the number of electrons it holds determines how the element will react with other elements.

GROUP

PERIOD	1	2	3	4	5	6	7	8	9	10	11	12	13	14	15	16	17	18
1	1 H																	2 He
2	3 Li	4 Be											5 B	6 C	7 N	8 O	9 F	10 Ne
3	11 Na	12 Mg											13 Al	14 Si	15 P	16 S	17 Cl	18 Ar
4	19 K	20 Ca	21 Sc	22 Ti	23 V	24 Cr	25 Mn	26 Fe	27 Co	28 Ni	29 Cu	30 Zn	31 Ga	32 Ge	33 As	34 Se	35 Br	36 Kr
5	37 Rb	38 Sr	39 Y	40 Zr	41 Nb	42 Mo	43 Tc	44 Ru	45 Rh	46 Pd	47 Ag	48 Cd	49 In	50 Sn	51 Sb	52 Te	53 I	54 Xe
6	55 Cs	56 Ba	57-71	72 Hf	73 Ta	74 W	75 Re	76 Os	77 Ir	78 Pt	79 Au	80 Hg	81 Tl	82 Pb	83 Bi	84 Po	85 At	86 Rn
7	87 Fr	88 Ra	89-103	104 Rf	105 Db	106 Sg	107 Bh	108 Hs	109 Mt	110 Ds	111 Rg	112 Cn	113 Uut	114 Fl	115 Uup	116 Lv	117 Uus	118 Uuo

Lanthanides	57 La	58 Ce	59 Pr	60 Nd	61 Pm	62 Sm	63 Eu	64 Gd	65 Tb	66 Dy	67 Ho	68 Er	69 Tm	70 Yb	71 Lu
Actinides	89 Ac	90 Th	91 Pa	92 U	93 Np	94 Pu	95 Am	96 Cm	97 Bk	98 Cf	99 Es	100 Fm	101 Md	102 No	103 Lr

Credit for the periodic table goes to the Russian chemist and schoolteacher Dmitry Mendeleyev, who published his work in 1869. He was looking for a better way to arrange the sixty-three elements known at the time. And the information on some of the chemicals was wrong. We now know there are well over one hundred.

Mendeleyev got the idea from the way a deck of cards is sorted by suit and value. It was like working on a giant jigsaw puzzle with a third of the pieces missing and some of the pieces he did have bent. The genius of Mendeleyev was that he left room on the periodic table for elements not discovered. He correctly predicted that more would follow and even the properties some of them would have. Indeed, several were discovered during his lifetime and had the properties he had predicted.

191. I've heard of hard water, but what is heavy water?

The term "hard water" generally refers to water that contains a lot of dissolved minerals. "Soft water" has many of those minerals removed. The minerals in hard water give water its fine taste, but those same minerals make it more difficult for soap to do its job.

"Heavy water" is a nickname for water whose hydrogen atoms are really deuterium, an isotope of hydrogen. Atoms that have the same number of protons but different numbers of neutrons are called isotopes. Most hydrogen atoms have one positive proton in the nucleus and one negative electron going around it, but a few hydrogen atoms have both a proton and a neutron making up their nucleus: that's deuterium, abbreviated ^2H (for hydrogen with an atomic weight of 2) or D. Because of this additional particle, atoms of deuterium are slightly heavier than normal hydrogen atoms and molecules of water made up of deuterium and oxygen (D_2O) instead of hydrogen and oxygen (H_2O) are slightly heavier than normal water molecules. So we call water with deuterium "heavy water." Heavy water is rare; there is one molecule of heavy water for every three thousand molecules of "ordinary" water.

Deuterium has cropped up in all sorts of places. Most of the world's supply of heavy water is found in Norway and Canada. Canada has some nuclear reactors that use deuterium as a moderator to slow down the speed of neutrons so they have a better chance of splitting an atom of uranium. Deuterium is not radioactive; we ingest some deuterium when we drink water.

Deuterium has an interesting history. It was discovered in 1931 by the American chemist Harold Urey. The Nazis tried to use deuterium as a moderator to make atomic bombs during World War II. Fortunately, they didn't get very far.

192. What is in shampoo that makes it clean your hair?

Each hair strand is rooted in a hair follicle that extends below the scalp. The hair follicle has a gland that makes an oil called sebum to moisten the hair follicle. The sebum prevents the skin and hair from having a dry and brittle quality.

Shampoos are basically soaps that eliminate excess oils and dirt that collect on the scalp. Water has a slight electrical charge, which causes its surface to act as a thin, rubberlike membrane; we call this surface tension. Soap breaks down the surface tension of water and allows it to get into tiny places, such as strands of hair or the fabric of our clothes. Many shampoos contain coconut oils and oleic acids that restore some of the hair's natural oil that shampoos consume, and most have added perfumes to make our hair smell nice.

1. Cut a boat shape about one inch wide and two inches long out of cardboard (piece of a cereal box).
2. Make a small notch at the back.
3. Fix a tiny piece of soap in the notch.
4. Put your boat in a bowl of water and watch it move.
5. The soap weakens the surface tension behind the boat, so the boat is pulled forward by the stronger surface tension in front.

7

The Strange World of the Atom

193. What is quantum physics?

Quantum physics, or quantum mechanics, is a branch of science that deals with how nature behaves at the atomic and subatomic levels, the realm of the infinitesimal.

The foundations of quantum mechanics were established in roughly the first forty years of the twentieth century, from about 1895 to 1935. The big names in this field are Niels Bohr, Max Planck, Albert Einstein, Werner Heisenberg, Max Born, John von Neumann, Paul Dirac and Wolfgang Pauli, Louis de Broglie, and Erwin Schrödinger, with a host of more minor players.

Max Planck stated that energy waves are radiated and absorbed in discrete, or separate, bundles, or "quanta," and that the amount of energy is related to the frequency (number of vibrations per second) of the waves. He was clever enough to come up with a formula that says so: E (energy) equals h (Planck's constant) times f (frequency), or $E = hf$.

We can kind of see quantum mechanics at work in everyday life. Heat up a piece of iron, say, a nail. Keep increasing the heat. The first color of the glow given off is red. Keep on heating it and soon it turns orange, then a tad blue, and finally it is white-hot. Red is the lowest frequency of visible light, so it requires the least heat energy. The glowing white-hot horseshoe is emitting all the colors and even ultraviolet, and that takes lots of energy.

There are three central ideas of quantum physics. First, energy is not continuous but comes in small, discrete units. Second, electrons behave both as waves and as particles. Last, movement of these particles is random and not predictable. One can't know both the position and the velocity of a particle at the same time.

The obvious question then is: What good is quantum mechanics? Much of our modern technology rests on quantum physics. Lasers, transistors, microchips, LEDs, MRIs, electron microscopes, USB flash drives, and superconductors all depend on quantum effects.

And finally, quantum theory settled a centuries-old debate about the nature of light. The Englishman Sir Isaac Newton claimed that light was a particle, but the Dutchman Christiaan Huygens showed that light is a wave. The quantum theory married these two seemingly competing ideas into one, stating that matter can behave as a wave and a wave can act as matter.

194. What happens when you split an atom?

Splitting atoms releases a tremendous amount of energy. This process is called fission. But not just any old atom will work. Fission requires the right kind of atom, and uranium, specifically uranium-235 (U-235), happens to be the atom of choice—it's the most important atom in the production of nuclear power and atomic bombs.

That "235" is the sum of neutrons and protons in the nucleus, or center, of the atom. Protons have a positive electrical charge, and neutrons have no charge. The number of protons in the U-235 nucleus is 92; this is matched by the same number of electrons, with a negative charge, that swirl in orbits around the nucleus. Subtract 92 from 235 and you get 143, which is the number of neutrons in the nucleus.

Uranium is a common element on Earth and has been around since the Earth was formed. The U-235 atom "decays," or changes (transmutes), into a less radioactive element all by itself by throwing off pieces of the nucleus—mostly alpha particles, which consist of two protons and two neutrons. Radioactivity is the streaming or emission of particles or rays from an atom's nucleus. Sometimes the term "nuclear decay" is employed. This natural radioactivity is happening all the time. A handful of U-235 atoms would take billions of years to completely change into lead, because there are a gazillion atoms in a handful—fairly close to size of our national debt!

But the fact that uranium-235 could undergo changes provided scientists with a tantalizing clue in the late 1930s. What would happen if a neutron hit a U-235 nucleus? Particles with the same charge, such as positive and positive or negative and negative, repel each other. So it has to be a neutron, because a particle with a positive charge (proton) would be repelled by the positive protons in the nucleus, and a particle with a negative charge (electron) would be repelled by the negative electrons already orbiting the nucleus.

Lise Meitner, Fritz Strassman, Otto Frisch, and Otto Hahn in Germany and Enrico Fermi in Italy are credited with first inducing radioactivity artificially. If a neutron strikes a uranium nucleus at high speed, it goes right though the nucleus and comes out the other side. Nothing happens to the atom. But if a neutron can be slowed down sufficiently before it hits the uranium nucleus, it deforms the nucleus, which splits apart. Each part takes a share of the original atom's protons and electrons, giving rise to two atoms of different elements, typically barium and krypton.

Here's the wonderful part. Along with the two new elements, at least two neutrons are released. These neutrons can each split two other uranium atoms. And when these atoms split, a minimum of four neutrons split four more uranium atoms. As this cascading effect continues through a mass of uranium, we have a chain reaction. This process continues until all the U-235 atoms are split.

This is what happens in an atomic bomb, which is essentially a runaway nuclear chain reaction (see page 271). In the fission process, a tiny bit of mass is lost, and that lost mass goes into pure energy of blast, heat, and light. The equation that explains the equivalence between mass and energy is Albert Einstein's famous "energy equals mass times the speed of light squared" ($E = mc^2$). E is energy in joules, m is mass in kilograms, and c is the speed of light in meters per second (see page 273).

Nuclear power plants and nuclear aircraft carriers and submarines use uranium-235, but with a controlled chain reaction, which slows down the rate of the splitting of atoms tremendously.

There is no way to use natural uranium for a bomb. Natural uranium is mostly uranium-238; only 0.7 percent is U-235. That's another way of saying that one in 140 uranium atoms are of the U-235 variety. The uranium used in power plants is enriched to 3 percent; that is, three in every one hundred atoms is U-235. The uranium used in a bomb must be enriched to about 90 percent U-235. So a nuclear power plant cannot blow up like a nuclear bomb. However, the chain reaction can "get away" from a nuclear power plant; witness the Three Mile Island and Chernobyl meltdowns.

Two commercial methods currently exist for enriching uranium: gas centrifuging and gaseous diffusion. Both require large-scale production facilities. Both also require wads of money and are time-consuming and tedious. The gas centrifuge method uses a series of rotating cylinders in which the heavier U-238 atoms are slung to the outside and the lighter U-235 atoms collect near the center of rotation. It is a fancy version of the old-time farmer's cream separator. The gaseous diffusion process involves mixing the raw uranium with fluorine gas to get the gaseous compound uranium hexafluoride. This gas bombards screens with tiny holes in them. The smaller U-235 atoms get through, while the larger U-238 atoms do not. This process is repeated thousands of times for any one sample, and each time the uranium is a bit more U-235 and a bit less U-238.

If you have been following the news, you will have noted that Iran is enriching uranium. Their leaders say it is for peaceful purposes, but Western governments consider their intentions suspect; the plutonium from the spent fuel rods of a nuclear power plant can be processed and fashioned into a bomb. Plutonium-239 is one of the by-products of fission in a nuclear power plant. Pu-239 can be chemically separated from the spent fuel rods. Pu-239 is a good choice for weapons builders and was used in the second bomb dropped on Japan in August 1945.

195. How do glow-in-the-dark objects work?

Phosphorescence is responsible for an object's glow-in-the-dark quality, but to fully grasp this concept, it would be helpful to first understand a related process, called fluorescence. Fluorescence is the emitting of light of a longer wavelength by an object when hit with light of a shorter wavelength. Simply put, a fluorescent object gives off light that we can see when it is struck by ultraviolet light, which we can't see, and the visible light is emitted immediately. Some gemstones, like rubies, emeralds, and diamonds, are fluorescent, as are black light posters. A phosphorescent object, on the other hand, emits light over a period of several minutes, even after a light source is taken away; hence the name "glow-in-the-dark." In most cases, glow-in-the-dark objects can be recharged again and again by exposure to regular light. The so-called regular light can come from sunlight, fluorescent lightbulbs, or incandescent light bulbs.

Keep in mind that those glow sticks that you snap and shake are not phosphorescent materials. Glow sticks, especially popular for Halloween trick-or-treaters, produce light by mixing two different chemicals, which react with each other to produce light (see page 193).

Ultraviolet light striking either fluorescent or phosphorescent materials boosts the electrons orbiting the nuclei of their atoms into orbits farther from the nuclei. It's something like putting the planet Mercury into Venus's orbit around the Sun. The atom is said to be in an excited state. When the electron goes back to its original orbit, the atom emits light, only now it is light we can see. In phosphorescent materials, the electrons take their sweet time, going back home to their original orbit a few billion at a time. So they give off light over a time of a few seconds to several hours, depending on the kind of material.

Dating back to the 1930s, zinc sulfide was employed extensively for its luminescent properties. Luminescence is the emitting of light not

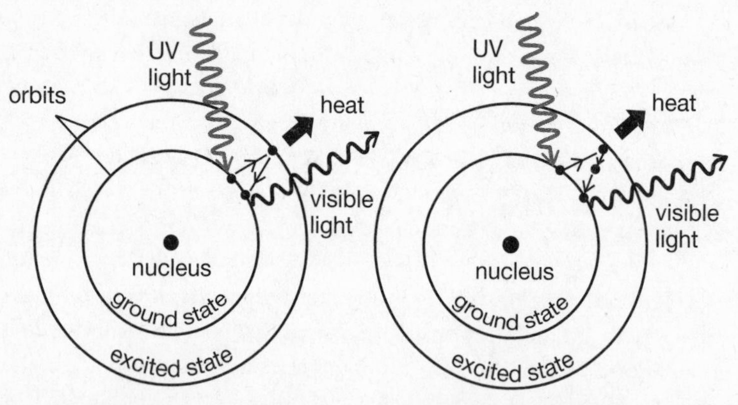

In fluorescence, the UV light delivers enough energy for the electron to move to an excited state, plus a little extra which is released as heat. The electron quickly falls back to the ground state and releases the remaining energy as visible light.

Phosphorescence works almost the same way, but the electron goes from its excited state to an intermediate energy level (by changing the direction it's spinning in). It can stay there a little while before releasing light and returning to its ground state.

resulting from heat. Zinc sulfide replaced the dangerous radioactive radium that was placed on watch dials and aircraft instruments so that they could be read in the dark. Later, strontium oxide aluminate was developed, which lasts ten times longer.

Today, such pigments are used on exit signs, pathway markings, and those decorative stars you can stick on ceilings. Timex, the watch company, patented a system they called Indiglo some years ago, which uses a transformer that takes 1.5 volts from the watch battery and multiplies it by a factor of ten, to 150 volts that excite a thin layer of phosphors.

It was the study of phosphorescent materials that led to the discovery of radioactivity in 1896 by the Frenchman Henri Becquerel. The unit for radioactivity, the becquerel (Bq), is named after him. Both the Moon and Mars have craters designated Becquerel in his honor.

196. Can you really use your tongue to tell if a battery is dead or not?

A good nine-volt battery has a high enough voltage to produce a mild shock if you briefly touch the contacts together with your tongue. You might also get a nasty metallic taste in your mouth.

If the battery is completely run-down or depleted, little or no shock would be experienced by the "tester." That's the idea or theory of using the tongue as a battery tester. Read on to discover that the tongue procedure is not the best idea in the world.

The other commonly used batteries, AA, AAA, C, and D cells, are all 1.5 volts. So they most likely will not have sufficient voltage to produce this "tongue shock." A nine-volt battery is actually six 1.5-volt AAA cells wired in series. Voltages over thirty volts may be dangerous.

It may work better if you put one terminal of the battery to your lip and the other end to your tongue. The low voltage and current will not hurt you. The voltage is the electrical force or pressure pushing the electrons (electricity) on a conductor or wire. Current is the flow of the electrons on that same conductor or wire. Voltage is analogous to the water pressure in your plumbing pipes. Current is akin to the water flowing in the pipes.

The body's resistance to electricity is primarily concentrated in the skin and varies with the skin's condition. Dry skin has a lot of resistance, so touching a nine-volt battery or a twelve-volt car battery will not produce a shock, because there's not enough current flowing to overcome that resistance. But if the skin is wet or punctured from a cut, abrasion, or needle, the resistance of the skin goes way down, and more current flows, which raises the likelihood of a shock. Under no circumstances should you touch a battery of any kind using both hands if you have an open wound, sore, cut, etc.; the opening gives the current access to wet and saltier tissue, which transmits it better. You don't want to put any voltage across your heart from one hand through your body, to the other hand.

The perception of an electric shock is complex. It depends on the voltage, duration, current, and frequency of the electricity and the path it takes. Current entering the hand needs to be about five to ten milliamperes (mA) for a person to feel a shock; currents of under 100 mA can be deadly depending on what parts of the body they pass through.

But let me suggest that there are better ways to check a battery. Discount stores, hardware stores, and auto supply stores sell those cheap little multimeters, or battery testers, for not much more than five dollars. Take care of yours, and it will last a lifetime. Plus you can use the multimeter to measure other voltages, find electrical shorts and open circuits, and measure current flows. In addition, if you have one of those multimeters lying around, visitors will be impressed with your electrical knowledge and handyperson skills. So it makes a good conversation piece.

When I was a kid growing up on that Seneca farm, there were young lads who would test an electric fence in a decidedly different fashion. Instead of touching the fence, they would walk up to the electric fence, pull their trousers down, aim, and . . . Not a good idea!

197. Do cell phones give off radiation? Can you get cancer from using one too much?

The jury is out on that question. The issue came to national attention in 1993, when a Florida man appeared on a national talk show and claimed his wife's brain tumor was caused by radio frequency radiation from her cell phone. His lawsuit was dismissed in 1995 due to a lack of scientific and medical evidence.

There is some anecdotal support, but not scientific evidence, of a link between brain tumors and extreme cell phone use. The lawyer who defended O. J. Simpson, Johnnie Cochran, developed a brain tumor on the side of his head that he held his cell phone against. Senator Edward Kennedy of Massachusetts was diagnosed with a brain tumor in May 2008 and died in August 2009. Some believe his brain tumor was caused by heavy cell phone use. One has to be careful about these kinds of stories. Cause and effect can be quite tricky.

Cell phones use radio frequencies (RFs) that are higher than the frequencies of radio and TV waves but somewhat lower than those of radars and microwave ovens. The amount of that RF energy a person using a cell phone is exposed to varies with the distance from the cell tower, the frequency of cell phone use, the cumulative length of their calls, and the age of the cell phone. The older, analog cell phones emit more radiation than the newer digital ones. The radio waves used by cell phones are nonionizing, unlike the radiation from X-rays and radioactive materials, which is ionizing. Ionizing radiation has enough energy to knock an electron out of its orbit around an atom's nucleus, which causes the atom to become charged, or ionized (see page 281).

The evidence so far indicates that the energy from a cell phone does not contain enough energy to damage genetic material, which could lead to cancer. But the problem is that cell phone use is relatively new. No one knows for sure about long-term use, say, over a lifetime. Scientists have not had the opportunity to carry out long-term studies. Brain tumors often take twenty or more years of exposure to develop, and since most people began using cellphones about twenty years ago, we're just starting to be able to see long-term effects.

Nor do scientists know enough about frequent use, when people are on their cell phone for several hours a day. Few authoritative people claim that there is a definite cause and effect between brain cancer and cell phone use. But prudent people in the field are saying, "better safe than sorry."

The European Union has been doing a massive study. European countries have been using cell phones since the early 1990s, so they have a longer timeframe in which to study the problem. The study was completed in 2012. The report indicates that there is no link between cell phone use and cancer, but the information has not satisfied all critics.

198. Why is radiation used in treating cancer?

Radiation treatment of cancer, called radiation therapy or radiotherapy, comes in two forms. One kind of treatment uses beams that come from machines, much like X-rays but at much higher energy levels. Indeed, X-rays themselves are used for cancer treatment. The second method uses waves or particles that come from radioactive sources.

Cancer is cell division gone wild, or out of control. The idea of radiation therapy is to kill cancer cells. Radiation ionizes atoms and molecules, which means that it knocks electrons out of orbit around the nucleus of the atoms. The atoms are no longer neutral; they have a charge of positive or negative. This ionization affects the nuclei of the cells, and particularly the DNA in the nuclei, which influences the cells' ability to grow and divide. The hope is that many of the cancer cells will fail to divide and therefore will die. Good cells are also damaged during radiotherapy, but those cells have a better chance of repairing the damage to their DNA and recuperating. Healthy cells rejuvenate on their own.

In addition, doctors restrict the amount of radiation to what is known to be safe for the normal cells. They can also target the therapeutic dose to the site of the cancer so as to minimize the exposure of normal tissues.

In many cases, the radiation therapy is delivered by an external beam aimed at the cancer site, as in the case of the gamma-ray knife, used for brain tumors. A rigid head frame is attached to the patient's skull so the head cannot move during therapy. Over two hundred beams of low-intensity gamma rays from cobalt-60 converge on a deep-seated brain tumor, adding up to an intense dose of radiation delivered to a small area safely. Treatment can last anywhere from minutes to several hours.

The second method of radiotherapy is to introduce radioactive sources into the body. These may take the form of pellets, or seeds, in or next to the targeted cancer sites. Prostate cancers, for example, can be treated by implanting radioactive seeds or pellets directly into a localized cancer site. It is an outpatient procedure and has a short recovery time. Another approach is to inject or have the patient swallow a radioactive substance that naturally binds to the type of tissue where the cancer arose or has an antibody attached that targets such tissue.

Radiation therapy is often used in combination with surgery (sometimes before it, to reduce the size of a tumor), chemotherapy, and/or hormone therapy, because it can decrease the chances of cancer's recurring. And even when a cancer cure is not possible, radiation therapy is used to relieve pressure or reduce pain.

Radiotherapy does have side effects. Tissue damage, hair loss, fatigue, infertility, and nausea are some of the most common.

199. What makes an object transparent?

We've all noticed that most liquids and gases are transparent, or clear, and that most solids are opaque, meaning that light can't pass through them. This is a fundamental difference between solids and liquids.

Solids are materials in which the molecules are held tightly together by bonds that give the matter rigidity. When a solid melts, the strength of those bonds lessens and the molecules begin to align themselves randomly. This movement from ordered to random is the reason that light can pass through liquids and gases. The molecules are not stacked neatly together, and the gaps and holes allow light to pass through (with a few exceptions, like liquid mercury, which is opaque). Think in terms of a brick wall. When the bricks are neatly laid and mortared together, no light gets through. But if bricks are lying in a random pile, light can get through the gaps.

But that doesn't explain how light gets through glass, which, of course, is a solid. Now we have to go to the atomic level. When light hits a pane of glass, it sets up vibrations in the atoms of the glass. A system tends to oscillate, or vibrate, with a larger amplitude at some rates of vibration (frequencies) than at others. The vibration rate at which a substance responds most strongly is referred to as its resonant frequency, or natural rate of vibration. A bell rings at a certain tone, or frequency. A tuning fork vibrates at a set frequency. So do the electrons in different kinds of matter.

Light in the visible region hitting an atom in glass forces its electrons into vibration. The atom holds the energy for a bit of time, then passes it on to the next atom, causing that one to vibrate a tad and then pass it on to the next, and so on, until finally the light comes out of the glass at the same frequency as the light that went in.

In a material, such as transparent glass, the vibration rate of the atoms is such that the energy is transferred from atom to atom, so that a piece of light hitting the first atoms are able to cause the last atoms to emit light. In a solid material, such as steel, which does not transmit light, the incoming light causes the first few atoms to vibrate a bit, but that vibration dies down quickly because it cannot pass its vibration to the next atoms. Instead, the light energy heats up the material.

This process of absorptions and re-emissions of the light energy through glass takes time. So while light travels through space or air at about 186,000 miles per second, it travels through glass at about 124,000 miles per second.

If ultraviolet light hits glass, there is a strong resonance between the electrons and nuclei of the glass atoms, which causes them to vibrate violently. The atoms collide repeatedly with nearby atoms and give up energy as heat. This transforms nearly all the energy of the ultraviolet light wave to heat energy; very little is left in the form of a light wave to pass through the glass. Infrared light waves, which are much longer than those of visible light and have a lower frequency, vibrate not only the electrons but the entire structure of glass. This vibration increases the internal energy of the glass and makes it warm. The infrared light can't get through. That is why our car heats up when we leave it sitting in the

Sun. Visible light gets through the glass, but the infrared, with its longer waves, can't get back out and the interior heats up. It is the classic greenhouse effect. In short, glass is generally transparent to visible light but not to ultraviolet or infrared.

200. How do atomic bombs work?

There are two types: fission and fusion. The fission type of atomic bomb works by splitting an atom. The fusion kind brings together, or fuses, lighter atoms. Either type, the fission atom bomb (A-bomb) or the fusion hydrogen bomb (H-bomb), releases a tremendous amount of energy.

A fission bomb uses uranium-235 (92 protons and 143 neutrons) or plutonium-239 (94 protons and 145 neutrons) to create a nuclear explosion (see page 276). A neutron splits the U-235 atom's nucleus into two fragments, with the release of an enormous amount of energy and a loss of two or three neutrons in the process. The released energy takes the form of kinetic energy, heat, and light. Meanwhile, the two or three neutrons given off can then split other U-235 atoms, which give off more neutrons, which split more atoms, and so on—a chain reaction.

Energy is released by splitting the nucleus because the fission fragments and the neutrons together weigh less, or have less mass, than the original U-235 nucleus. The lost mass, or mass defect, is actually turned into pure energy, which follows Einstein's famous equation, $E = mc^2$. Because c stands for the speed of light and c squared is a huge number, just a little bit of mass is equivalent to a fantastic amount of energy. A pound of uranium, for example, has the same energy as three hundred thousand gallons of gasoline. The critical mass is the minimum mass of fissionable material needed to sustain a chain reaction. Reports indicate that as little as thirty pounds of U-235 or twelve pounds of Pu-239 can be fashioned into an atomic bomb. In any fission bomb, the nuclear material must be kept in separate subcritical masses that don't, on their own, support fission.

There are two general methods of making an A-bomb: gun and implosion. The first A-bomb, used on Hiroshima on August 6, 1945, and

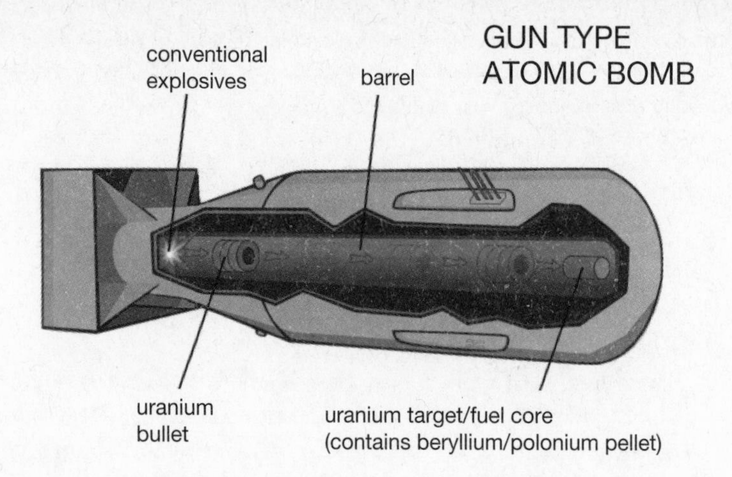

dubbed Little Boy, was the gun-type, while the one used on Nagasaki on August 9, 1945, and named Fat Man, was the implosion kind.

In the gun type, two subcritical masses of enriched U-235 are kept separate from each other in a sealed cylinder like the barrel of a gun until the time of detonation, when explosives propel one piece down the barrel like a bullet into the other piece. Neutrons, needed to start the chain reaction, are produced by using a small pellet of polonium and beryllium within the fuel core. A piece of foil keeps the two elements separated. When the masses of uranium come together, the foil is broken, and the polonium emits alpha particles, which collide with the beryllium to produce the neutrons needed to initiate the chain reaction.

Because the gun type is necessarily long, the first bomb, Little Boy, looks like an overgrown cigar. Scientists were so sure it would work back in 1945 that they never tested it beforehand. At the time, enriched uranium-235 was so scarce, there wasn't enough to drop another Little Boy–type bomb for many months.

The implosion type weapon is one in which the fissile Pu-239 is surrounded by high explosives that compress the material sufficiently to make it detonate. It is similar to packing loose snow into a snowball.

Some excellent movies have been made about the making of the atomic bomb. *Day One* is a 1989 TV movie with big-name stars like Brian Dennehy, Hume Cronyn, and Hal Holbrook. *Fat Man and Little Boy* is a big-screen movie, also made in 1989, with Paul Newman playing General Leslie Groves.

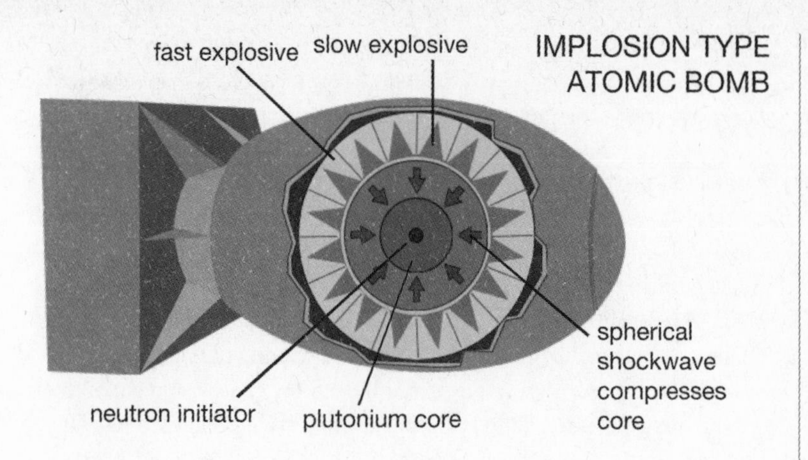

201. What does Einstein's equation, $E = mc^2$, have to do with atomic bombs?

Not a whole lot, really. Albert Einstein wrote his famous equation practically as an afterthought to his 1905 theory of relativity. He discovered that there is an intimate relationship between mass and energy. Energy is the ability to do work, and mass is the amount of substance of something—how much stuff there is. They may seem very different, but mass and energy are interchangeable aspects of the same thing. Scientists refer to this connected unit as mass-energy. The equation tells us that even a tiny bit of mass contains a huge amount of energy. The letter *m* stands for mass and the letter *c* for the speed of light. The speed of light is a huge number (186,000 miles per second), and its square is even huger. Mass multiplied by the square of the speed of light yields an unbelievably large figure. So even a tiny bit of mass holds a tremendous amount of energy.

We don't experience this in everyday life. We don't see that every gram of water or soil or soap contains the energy to run whole cities. It's like a rich person who doesn't spend any money; no one can tell how rich they are. In everyday life, our energy-producing activities are chemical processes, from metabolizing food to burning wood in a fireplace, gasoline in an engine, or coal in a power plant.

Want energy? The name of the game is to lose mass. Chemical processes do cause a loss of mass, but it is so tiny it can hardly be measured. That's why we don't get much energy from chemical processes, even dynamite, or TNT. A pound (454 grams) of TNT exploding loses only half a billionth of a gram of mass.

Nuclear processes are quite different from chemical processes. Chemical reactions involve the transfer, loss, gain, and sharing of electrons. Nothing takes place in the nucleus of the atom in chemical reactions. "Nuclear" means working with the inner core of an atom, the nucleus, rather than with the electrons going around the nucleus. Nuclei of uranium atoms, for example, can split into two pieces, and when they do, the two pieces don't add up to the original whole. That seems odd at first and defies commonsense thinking. Cut a loaf of bread into two pieces, for instance, weigh the two pieces on a scale, and you will find the two pieces add up to the weight of the whole loaf that you started with.

That does not happen at the nuclear level. If an atom of uranium is split, the two pieces together weigh less than the whole uranium atom, by about a tenth of 1 percent. The lost mass turns into pure energy. How much energy? Einstein's $E = mc^2$ tells us how much. A little bit of mass turns into an enormous quantity of energy.

Split a uranium-235 atom, then particles (neutrons) from that atom split two more atoms, which in turn split four more atoms, then eight more atoms, and you have a chain reaction (see page 261). If that happens in a split second, it's an atomic bomb. If you slow down that chain reaction and control it, it's a nuclear power plant. A nuclear power plant cannot explode like an atomic bomb. The fuel is different. Most reactors use uranium enriched to 3 percent U-235. An atomic bomb is uranium enriched to 90 percent U-235. The Iranian government is accused of trying to do such enrichment.

Isotopes are two or more forms of the same atom that have the same number of protons but a different number of neutrons. For example, uranium-235 and uranium-238 both have the same number of protons, namely 92. But U-235 has 143 neutrons and U-238 has 146 neutrons.

So Einstein's famous equation is not needed to build an atomic bomb. However, the equation does measure the size of the blast.

202. Why do doctors give you lead shields when you're getting X-rays?

Superman actually provides a good example of the effect of lead on X-rays. Superman, Clark Kent, was "more powerful than a locomotive," "able to leap tall buildings in a single bound," and "faster than a speeding bullet." And with his X-ray vision, he could see through solid objects, but not lead.

There is good physics behind Jerry Siegel and Joe Shuster's 1930s cartoon character. X-rays are a form of electromagnetic radiation, just like light, microwave, cell phone, radio, and television waves. All these energy waves are bouncing up and down and sideways as they all move at the speed of light, and X-ray waves do this at a higher frequency than the rest. The higher the frequency (vibrations per second), the higher the radiation energy. The higher the energy, the deeper a ray will penetrate matter, which is the reason why dentists and doctors use X-rays to see images of our teeth and our bones. The X-rays go right through our flesh, but some are absorbed by the denser bones and cast a nice diagnostic shadow on a piece of photographic film.

That's the good news. The bad news is that X-rays, as well as gamma rays, are ionizing radiation. They knock electrons off atoms, leaving behind an ion. An ion is an atom that is missing one or more electrons. These atoms are not playing with a full hand, so to speak. They can change body chemistry in harmful ways. One such bad outcome is cancer.

So what can stop X-rays? Anything that has a lot of atoms with a lot of electrons going around the nuclei, because each time an X-ray knocks an electron out of orbit, it loses energy. The law of conservation of energy is in play here: If the atom containing the electron gains energy, then the X-ray loses energy.

To make your own X-ray shield, you'd need to get a material that has the most electrons per atom and the most atoms per cubic inch possible—something densely packed. Uranium is an excellent choice, because it has ninety-two electrons per atom and is nineteen times as dense as water. Gold works, too: seventy-nine electrons per atom and nineteen times denser than water. Or platinum, which has seventy-eight electrons per atom and is twenty-one times denser than water. But uranium, gold, and platinum are too expensive, and not much better than lead. And that's why we settle for lead, which has eighty-two electrons per atom and is eleven times denser than water. Lead is cheap, about one dollar a pound.

The X-ray technician drapes a lead-lined vest over the patient at the dental office. Then they stand behind a lead-lined wall and look through a lead-laced window before zapping the patient's mouth with X-rays. Because of its density of electrons, lead can stop any kind of ionizing radiation, so we find it used in all sorts of medical settings.

203. What is plutonium, and why is it used in bombs?

Plutonium is a radioactive metal. Its symbol is Pu, and it is named after the former planet Pluto. Plutonium has ninety-four protons in the nucleus, so we say it has an atomic number of ninety-four. Its most important isotope is Pu-239, which has 145 neutrons in the nucleus. Add those 94 protons to the 145 neutrons and you have an atomic mass, or atomic weight, of 239.

Pu-239 is valuable as a fuel in nuclear power plants and in atomic-bomb making.

Plutonium has a half-life of 24,100 years. If you start with a pound of plutonium, you will have a half-pound of plutonium left after 24,100 years. The other half-pound will have changed (transmuted) to some other element.

Plutonium is silvery white in pure form, but turns yellowish and greenish when oxygen hits it. It's very heavy stuff, having about the same density as gold. When a plutonium nucleus fissions, or splits, it releases a tremendous amount of energy in the form of kinetic energy and heat.

Virtually no plutonium exists in nature. In 1940, a team of scientists led by Dr. Glenn T. Seaborg and Dr. Edwin McMillan bombarded uranium with neutrons to produce neptunium, which transmuted to plutonium. During the Manhattan Project, large reactors were set up at Hanford, Washington, along the Columbia River, to produce plutonium to build an atomic bomb. Other Manhattan Project scientists at Los Alamos, New Mexico, discovered that plutonium fissions too quickly to be used in a gun-type weapon. So they designed an implosion-type weapon, one with a symmetrical shock wave that would compress the plutonium into an explosive critical mass. It's something like taking a handful of loose snow and squeezing it into a snowball. Plutonium pieces are arranged symmetrically around a beryllium-polonium core trigger. Explosives are packed on the outside, and, upon detonation, the explosives create a shock wave that compresses the Pu-239 onto the core, fission begins, and the bomb explodes. The whole fission reaction occurs in about six hundred-billionths of a second.

The first atomic bomb using plutonium was tested at Trinity Site in New Mexico in July 1945. A plutonium-239 implosion-type atomic bomb, named Fat Man, was dropped on Nagasaki on August 9, 1945.

Terrible as they were, this bomb and Little Boy, dropped on Hiroshima, arguably shortened the war, saved thousands, if not millions, of lives on both sides, and kept the Russians out of the area. An invasion of Japan, scheduled for November 1945, would have been so very bloody and devastating.

It was revealed after the war that the Japanese High Command had hoped to make the American invasion so costly in men and material that they could sue for peace and maintain their islands, government, and military infrastructure.

Not all plutonium is associated with atomic bombs. Another isotope of plutonium is Pu-238, which has a half-life of eighty-eight years. It emits alpha particles, so it is valuable as an electric power source for our space probes sent to the Moon, Jupiter, Saturn, and Mars. Artificial heart pacemakers using plutonium-238 have been around for about forty years. The NASA robotic rover to Mars, *Curiosity*, carried 4.8 kilograms, or about 10.5 pounds of Pu-238. *Curiosity* landed on the Red Planet in August 2012.

204. What is an atomic clock?

When you buy a clock or watch labeled "atomic clock," you are buying one that you can synchronize to the United States official atomic clock in Colorado. These clocks and wristwatches can pick up the shortwave radio transmissions of broadcast stations on several frequencies. The National Institute of Standards and Technology (NIST) and the US Naval Observatory air time signals from powerful transmitters on radio stations WWV, and WWVB from Fort Collins, Colorado, and WWVH, from Kauai, Hawaii. The frequencies are 2.5, 5, 10, 15, and 20 MHz. Reception of any radio signal can be affected by weather, location, time of day, time of year, atmospheric and ionospheric conditions, and other factors. Using so many frequencies ensures reception anywhere in the world for at least part of the day.

The radio receiver is tiny, about the size of the tip of a pencil, and can easily be embedded in computer chips and used in GPS units and cell phones. This miniature receiver need only pick up the radio transmission once every few days to correct any error in the receiver's clock. Generally, clock updating takes place at night, when the signal is strongest.

Many automated home devices, including computers connected to the Internet, are continually and automatically updated to the correct time. For example, Windows has a built-in service that allows your computer to reference the atomic clocks operated by NIST. Your current computer time is compared with the current atomic time and an adjustment is made to keep your local computer up-to-date with the exact time. Apple computers have a similar arrangement.

WWVB sends a digital time code on 60 kilohertz (kHz) that modulates, or changes, the power of the carrier signal picked up by the receiving antenna to indicate a 1, a 0, or a separator. The receiver decodes these bits to get indicators for the time, the day of the year, standard or daylight saving time, and regular or leap year. These indicators report what is called Coordinated Universal Time.

A grandfather clock keeps time by the steady back-and-forth oscillations of a pendulum powered by falling weights. Old-time clocks use a balance wheel with alternating backward and forward motion powered by a wound spring. Quartz watches depend on the vibrating oscillations of a crystal. The tiny crystal looks like a tuning fork, and will vibrate at a fixed rate (usually 32,768 times per second) when powered by a small battery.

Atomic clocks, the kind used by NIST, go back to 1945, when physicist Isidor Rabi determined that atoms maintain a steady, unchanging vibration rate even more precise than the pendulum, balance wheel, and quartz crystal.

Early atomic clocks used the vibration of ammonia molecules, but today's atomic clocks go with cesium. Every atom has several characteristic oscillation frequencies. To turn the cesium resonance into an atomic clock, it is necessary to measure one of the resonant frequencies accurately. This is done by locking a crystal oscillator to the principal microwave resonance of the cesium atom. This signal is in the microwave range of the radio spectrum, and just happens to be at the same sort of frequency as direct broadcast signals.

Estimates vary as to the accuracy of atomic clocks. One scientist claims precision to within one second in 126 years. Another report puts it at one second in 30 million years. Either one will get you to the church on time!

NIST has the big expensive atomic clocks that generate extremely accurate time signals, and you and I can buy inexpensive atomic clocks synchronized to the government's atomic clocks. We just have to set them to the right time zone.

Listen in to WWV on any one of the frequencies listed on page 278 on any shortwave radio receiver. The 15- and 20-MHz signals are strongest during the day. The 5- and 10-MHz come in best at night. You hear the steady *tick*, *tick*, *tick*, and then a human voice announces the time on the minute. It's quite amazing!

205. If we can't see atoms, how do we know what they are made of?

No one has ever really seen an atom, and yes, we humans like to see something to believe in it. In order to see something, light has to hit it and reflect some of that light to our eyes. But an atom is thousands of times smaller than a wave of light. So light can't hit a single atom and bounce off it.

Most of what we know about atoms comes from having something, like an electron or an alpha particle, hit an atom and then watching where the particle goes after it bounces out of the atom. We can see where particles go by using a cloud chamber or by monitoring them when they hit a specially coated screen. A cloud chamber is a closed box containing a supersaturated vapor of alcohol. A charged particle will cause the vapor to condense on it, leaving a trail of tiny bubbles that are visible to an observer.

The story of the atom goes way back to the Greeks. Democritus (460–370 BC) first proposed the idea of an "atom," or "iota," the smallest unit of matter that maintains the identity of the whole. Mendeleyev, with his periodic table of elements, showed in about 1869 that elements have regular repeating properties. John Dalton, an English school teacher, proposed the first modern atomic theory in 1808. In 1897, English scientist J. J. Thomson passed electricity through gases and discovered the electron. In 1911, Ernest Rutherford, a British physicist, shot alpha particles at a gold foil and proved that the atom has a dense central core, termed the nucleus. In 1932, James Chadwick discovered a particle within the nucleus, called the neutron. In the 1930s, particle accelerators, commonly referred to as atom smashers, were used to measure the sizes and masses of atoms and their parts. Today, the most accurate picture of the atom deals with quantum mechanics and discoveries made in the early 1900s by Niels Bohr, Erwin Schrödinger, Louis de Broglie, and Werner Heisenberg.

A new type of microscope was invented in 1981 called a scanning tunneling microscope (STM), which uses a tiny, sharp tip that conducts electricity and a piezoelectric scanning device attached to the tip. A piezoelectric device is a crystal that flexes when a voltage is applied and, conversely, generates a voltage when it is flexed. When the tip meets an atom, the flow of electrons, or current, between the atom and the tip changes. A computer registers the change in terms of x and y coordinates and maps the current's pattern over a surface. It is much like the old phonograph needle following the grooves of a vinyl record. This STM microscope allows scientists to see the outlines, but not the insides, of atoms.

206. What exactly is radiation, and what does it do to your body?

All electromagnetic waves, like light, radio, and television waves, are considered to be radiation. But when we hear the word "radiation," we generally think of nuclear radiation. There are three basic kinds of nuclear radiation: alpha, beta, and gamma.

An alpha particle is the nucleus of a helium atom. It consists of two protons and two neutrons. An alpha particle is big on a nuclear scale, so it cannot penetrate very far into matter; a piece of paper is enough to stop it. But alpha particles make changes in our bodies by knocking out electrons that orbit the nuclei in our cells (ionization); they're like bulls in a china shop. If alpha particles are kept outside the body, little harm can be done. If ingested, alpha particles are devastating. Read the account of the polonium-210 poisoning of former Russian spy Alexander Litvinenko to get a sense of how deadly alpha radiation can be if taken internally.

A beta particle is an electron, which is much smaller. Betas bore deeper into matter. Most beta particles can be stopped by thin pieces of plastic or metal. Beta particles penetrate deeper than alpha particles but are less damaging than alpha particles that have traveled equal distances. Alpha particles are huge compared to beta particles and cause havoc in tissue. If alpha particles are like bulls in a china shop, then beta particles are like poodles in a china shop.

Direct exposure of the skin to beta particles can damage and redden it. Inhaling or ingesting beta particles is of greater concern. Inside the body, beta particles can damage tissue and organs and lead to cancer later in life.

Gamma rays are similar to X-rays but have higher energy. Gamma radiation is very penetrating and is often used to treat tumors, but uncontrolled exposure can cause widespread damage in many body tissues. Because gamma rays have no mass, they are not easily stopped or slowed down by collisions with matter.

Radiation can be helpful or harmful, because it has the potential to either kill a person or cure a person. A single large dose can be fatal, but the same dose over a long period, say a month or a year, can have little

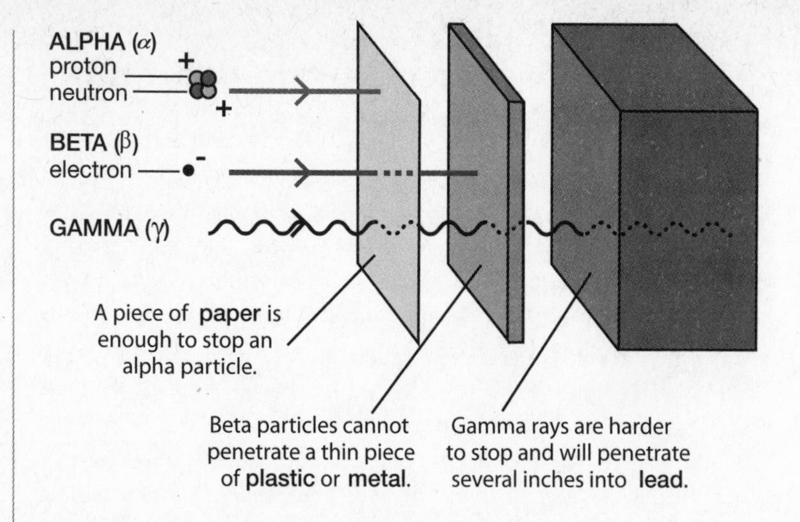

or no effect. A large dose of radiation distributed over the entire body can cause death, but the same dosage directed to a small area, say, a cancerous tumor, can be delivered safely.

The troubling question about radiation is: What are the effects of very low levels? Is there a threshold, or minimum value, of radiation that has an effect on the human body? Even one hundred years after the discovery of radiation, there is much debate on this topic.

207. How does static electricity make items cling together?

I know what you mean! I have three acrylic long-sleeve winter shirts. When these shirts come out of the clothes dryer, my black cotton socks have to be pulled off the shirts. A crackling sound can be heard, and sparks can be seen. This is static electricity, where the electrons are not moving on a conductor, such as a wire. Current flow electricity implies moving, not static, charges.

Clothes, of course, are ultimately made of atoms, and atoms consist of electrons, protons, and neutrons. Electrons have a negative electric charge, protons a positive charge, and neutrons no charge. An object that has an equal number of protons and electrons has no net electric charge. So it is neutral, that is, not charged, and does not cling to anything.

While tumbling in the clothes dryer, the acrylic shirts acquire a negative charge and the cotton socks get a positive charge. The charges are caused by friction between the two types of material. Opposite charges attract, so the sock clings to the shirt. When the sock is pulled from the shirt, the charge tries to equalize, or become neutral, by producing a tiny lightning bolt to discharge itself. This little lightning bolt, which see as a spark, can be as high as thirty thousand volts but holds hardly any current.

What type of charge a material acquires by friction is based on the triboelectric series, a list first published in 1757 on static charges. Materials are listed in order of the polarity of charge separation when they are touched with another object. A material toward the bottom of the series, when touched to a material near the top of the series, will attain a more negative charge. Acrylic shirts and cotton socks are sufficiently far apart on this list to ensure a very fine exchange of electrons, and a mighty fine spark.

8

Science Through the Ages

208. Why was 1905 called Einstein's extraordinary year?

In 1905, when Albert Einstein was twenty-six years old and a technical examiner in the Swiss patent office, he wrote four major papers that would forever influence science. The first, for which he later won the Nobel Prize, was a paper on the photoelectric effect. He described how light could behave as both a wave and a particle. The photoelectric theory laid the foundation for quantum physics.

The second paper proved the existence of atoms and molecules by noting that the jerky, apparently random motions of tiny particles seen in water, called Brownian motion, is caused by the impact of individual atoms.

The third paper is the one that made Einstein famous: it proposed his special theory of relativity. Einstein asked himself a simple question: "If I could travel on a light beam, what would I see?" He established that time and space would appear different for you and for a person not traveling with you. No matter what your speed, you will always measure the speed of light to be a constant, namely about 186,000 miles per second. As a person on a train travels near the speed of light, for example, time slows down for them relative to someone not on the moving train and the train is heavier (mass increase) and appears to be shorter (length contraction).

In his fourth paper, Einstein demonstrated the equivalence of matter and energy. He proved that mass and energy are two sides of the same coin. Even though Einstein never worked on or designed the atomic bomb, his famous $E = mc^2$ equation proved that one was possible. Forty years later, in 1945, the first atomic bomb was detonated in the desert in New Mexico.

What are the practical effects of Einstein's legacy? His first paper, on the photoelectric effect, gave rise to modern electronics, such as television, fiber optics, solar cells, and lasers. The Global Positioning System, or GPS, takes into account Einstein's special theory of relativity: clocks on GPS satellites have to be corrected because they do not run at the same speed as clocks here on Earth. The behavior and action of the Sun are explained by $E = mc^2$, as hydrogen is converted to helium and the four million tons of its mass that is lost every second is turned into pure energy.

Einstein was an inventor as well as a theoretician. He was granted twenty-five patents for such things as a hearing aid and a refrigerator that had no moving parts.

Another legacy will be his persona. He's the man with too large a head, walking around in rumpled clothing, his white hair unkempt, smoking a pipe, playing the violin, sailing in a small boat, and showing up at his naturalization ceremony in shoes but wearing no socks. Einstein is seen as a pure genius, a scientist laboring in relative isolation. He arguably had no equal in the twentieth century, and we are not likely to see his like in the twenty-first century.

209. Did Isaac Newton develop calculus?

Calculus is a powerful tool that helped put men on the Moon, provided "smart weapons" to our military, and made hospital MRIs possible. It allows us to solve science and engineering problems that ordinary algebra can't handle—specifically, problems involving rates of change. The development of calculus goes back to the Greek era of Democritus and Archimedes. They were using terms such as "an infinite number of cross-sections" and "limits." They recognized the requirement to compute the areas and volumes of regions and solids by breaking them up into an infinite number of recognizable shapes.

Seventeenth-century giants in math and science, such as René Descartes, Pierre de Fermat, and Blaise Pascal, added much to the development of calculus. Formalization of calculus as a distinct branch of mathematics can be credited to Isaac Newton and Gottfried Leibniz.

Calculus is the math of change and motion. For Isaac Newton, it was change over distance and time. Gravity was the problem for Newton, and calculus was his instrument. He needed a higher form of mathematics to explain gravity, because the pull of gravity varies as the square of the distance between any two objects. So if you are twice as far away from the Earth, the pull of gravity is only one-fourth as much. For more than a century, a debate dragged on over who developed calculus, Newton or Leibniz. A bitter public feud developed between the two mathematical geniuses, with charges and countercharges of plagiarism. Today, both Newton and Leibniz are given credit for the development of calculus.

Newton, born on Christmas Day in 1642, is considered the greatest scientist who ever lived. When the bubonic plague swept through England in 1665, the schools closed and the young teacher returned to the Newton family farm. In a period of eighteen months, he discovered the laws of motion and universal gravitation, found the laws that govern light and color, and formulated a new math, later called calculus. He developed the reflecting telescope in 1668. His book *Principia Mathematica*, published in 1687, is the foundation of the physical sciences and considered the most influential science book ever written. The unit of force bears Newton's name. He was the first scientist ever knighted by royalty. His image appeared on the British one-pound note from 1978 to 1988, an honor he shares with the Duke of Wellington and William Shakespeare. Isaac Newton is buried in Westminster Abbey, a place generally reserved for kings and queens.

210. When and how did *E. coli* develop?

Escherichia coli, better known as *E. coli*, is a normal bacterium found in the lower digestive system of mammals. But some strains cause disease. We hear of outbreaks periodically, and several deaths each year.

The *E. coli* bacterium was discovered in human feces in 1885 by a German doctor, Theodor Escherich, and the genus is named for him; the species name, *coli*, means that it inhabits the colon. We actually need *E. coli* to develop properly; these bacteria supply many needed vitamins, such as vitamin B complex and vitamin K. *E. coli* is often in the news as a food-borne pathogen, but the vast majority of *E. coli* strains are harmless. About 0.1 percent of all bacteria in our intestine are *E. coli*.

Immediately after we are born, we acquire all kinds of bacteria, which live symbiotically with us—we help them to live and they help us to live. Just as there are human beings who are not very nice, indeed downright dangerous, so there are different strains of *E. coli* bacteria that can harm us.

At some point, an *E. coli* cell was infected by a virus. The virus had the ability to put its own DNA into the bacteria's chromosomes without harming the bacteria, and the DNA was then able to remain there. Every time the bacterium cell divides, the virus's DNA is passed on to every descendant cell, so now we have *E. coli* O157:H7, one of those dangerous strains. This strain of *E. coli* produces a toxin that causes severe damage to the cell wall of the intestine. The damage is so massive that blood vessels are destroyed, with nasty bleeding, or hemorrhaging. Symptoms usually start in three or four days, but they can begin anytime between one and ten days after eating the contaminated food. The person has cramps, becomes dehydrated, and experiences bloody diarrhea. The condition can be fatal for children, who can't tolerate much fluid and blood loss. Elderly people are also more vulnerable.

We can prevent *E. coli* infections by thoroughly cooking ground beef, not drinking unpasteurized milk and juices, rinsing produce thoroughly, keeping meat and its juices away from other foods while cooking, and washing our hands carefully in between handling them. Because *E. coli* lives in the intestines of animals, meat can become contaminated during slaughter. This can lead to infection in people who eat the beef; the risk is higher in ground beef, because meat from many different animals is mixed together, increasing the chance that one of them carries *E. coli*.

When these *E. coli* outbreaks occur, they create headline news. Most cases involve eating undercooked and contaminated ground beef. Most people recover without any medication. However, severe cases can lead to kidney damage and even death. And a disturbing new trend is the existence of some strains of E. *coli* becoming resistant to certain types of antibiotics.

211. When did we first discover the existence of another planet by telescope?

The five planets closest to the Sun, other than Earth, can be seen with the naked or unaided eye. So Mercury, Venus, Mars, Jupiter, and Saturn were known since the first time humans looked up to the heavens. These planets seemed to move about among the background stars. The word "planet" is rooted in the Greek word for "wanderer."

Uranus was the first planet to be discovered by telescope. In 1781, English astronomer William Herschel noted a "bigger than normal" star in the constellation of Gemini, the Twins. That big object turned out to be a planet.

Newton's laws allowed the prediction of the existence of Neptune even before the planet was discovered, making it the first planet to be discovered by mathematics. Something was tugging, or pulling, on the orbit of Uranus. Calculations done in 1846 showed that the gravitational effects of a nearby object perturbed the path of Uranus. Astronomers trained their telescopes on the exact spot where such a planet should be, and Neptune showed up!

Pluto was the first planet (although since demoted to "dwarf planet") discovered using a photograph. In February 1930, astronomer Clyde Tombaugh, working at the Lowell Observatory in Arizona, noticed that two images on different photographs had shifted slightly. From our distance, stars do not appear to move relative to each other—at least not over a period of weeks or months.

Pluto was "kicked out" of the family of nine planets in August 2006 by the International Astronomical Union that met in the Czech Republic. They placed Pluto in the family of dwarf planets, of which there are more than forty-four known. They said that Pluto does not dominate its moon Charon. Charon is half the size of Pluto and planets are much, much bigger than their companion moons. Also, Pluto has a highly elliptical orbit rather than the more sedate circular orbit enjoyed by the remaining eight "real" planets. Adding insult to injury, Pluto's orbit path is inclined steeply to the plane of the other planets, marking it as a true renegade.

Now astronomers can search for planets orbiting stars other than our Sun, or exoplanets. This is not an easy task. The light from stars is so bright that the glare overwhelms the light reflected from surrounding planets. It's like trying to see a birthday candle placed right in front of a searchlight. Astronomers have developed a few work-arounds to this problem. For example, they try to measure the gravitational effect of planets on stars. A planet tugs on the star as it orbits, causing the star's motion to change slightly. Also, if a planet's orbit takes it between its star and Earth, it will block some of the light, so astronomers can look for variations in the brightness of stars. Since 1990, over 880 planets have been discovered that orbit around other stars.

In 2014, NASA plans to launch the Terrestrial Planet Finder (TPF) an array of four optical telescopes and a combiner instrument. The light from the four telescopes will be combined in such a way as to cancel out the bright glare from the star. Hopefully, finding other planets will take a giant leap forward with this new technology.

212. Who or what built Stonehenge?

Stonehenge is an impressive site, almost mystical in nature. One of the most famous prehistoric sites in the world, Stonehenge sits atop a low hill and is surrounded by a near-treeless plain with cattle grazing nearby. Stonehenge, a ring of immense stones, is on the Salisbury Plain about one hundred miles west of London.

Stonehenge marks the seasons. At summer solstice, the Sun rises over the Heel Stone as seen from the Altar Stone. Construction started around 3000 BC, and the last known construction was around 1600 BC. The Heel Stone is a large block of sarsen stone outside the Stonehenge earthwork, close to the main road. It is sixteen feet tall and eight feet thick. The Altar Stone is seven feet tall and weighs an estimated six tons.

There is no general agreement on why Stonehenge was built. Some argue that it was a place of worship, or church, if you will. Others claim it was strictly an observatory to mark the seasons. Archeologists continue to study and ponder this magnificent structure.

Work went on by prehistoric people for a period of about fifteen hundred years. The twenty-five- to fifty-ton sarsen stones we see today came from quarries in Marlborough Down, a distance of twenty-five miles.

They were set upright in a circle and topped with a ring of stone lintels. The four-ton bluestones, placed during the third phase of construction, came from Wales, over 150 miles away. What an amazing feat of transportation!

My wife, Ann, and I visited Stonehenge in August 2006. Walking around the massive stones, I was thinking of those ancient people who built this edifice. What would their language sound like, what did they eat, how did they gather food, how did they stay warm, what kind of games did they play, and how did they govern themselves? And what induced these people to build anything on this lonely, windy plain near Salisbury?

There are some good books on the subject. One of the best is *Stonehenge: The Secret of the Solstice*, by George Terence Meaden. Another is *The Making of Stonehenge*, by Rodney Castleden. Yet another excellent work is *Stonehenge—A New Understanding*, by Mike Parker Pearson. The Internet is an excellent resource for some gorgeous pictures. If you ever have the opportunity to be in England, go see Stonehenge.

213. How did they figure out the speed of light?

Ole Rømer, a Danish astronomer, made the first estimate of the speed of light in 1676. Using a telescope, he observed the motion of Io, one of four large Galilean moons of Jupiter. Io would alternately move in front of Jupiter, and then go behind Jupiter. It took 42.5 hours for Io to go around Jupiter when the Earth was closest to Jupiter. He recorded that as Earth and Jupiter moved farther apart, Io's eclipse by Jupiter would come later than predicted. He estimated that it took twenty-two minutes for light to cross the diameter of the orbit of Earth. His calculation for the speed of light came out to 136,000 miles per second. Not bad for a first crude measurement. The accepted value today is about 186,000 miles per second.

Armand-Hippolyte-Louis Fizeau made the first successful measurement of the speed of light on Earth in 1849. He directed a beam of light at a mirror a few miles away. The beam passed through a rotating cog wheel. At a particular rate of rotation, the beam would pass through one gap on the way out and another on the way back. By knowing the distance to the mirror, the number of teeth in the cog wheel, and the rate of rotation, Fizeau could calculate the speed of light. He got a result of 195,732 miles per second, closer to the true value.

An American, Albert Michelson, measured the speed of light in 1926. He set up a rotating mirror system to measure the time for light to make a round trip from Mount Wilson to Mount San Antonio in southern California, a distance of about twenty-two miles. Michelson's precise measurements yielded a value of 186,285 miles per second, or 299,796 kilometers per second. He was awarded the Nobel Prize in 1907 for teaming up with Edward Morley, also an American, to try to measure the so-called ether.

Light is an electromagnetic wave and travels at the same speed as radio or television waves. Light takes 1.2 seconds to go from the Earth to the Moon. Sun to Earth time is 8.5 minutes. Sun to Pluto is 5.5 hours. The light of the nearest star is 4.3 years away, and light takes one hundred thousand years to cross our Milky Way Galaxy. We humans live in a very big house!

214. How did dinosaurs become extinct?

Dinosaurs roamed planet Earth for 160 million years; then suddenly they died off, 65.5 million years ago. There have been a half dozen prominent theories and as many less plausible theories to explain the dinosaur's sudden demise.

The reputable accounts over the past two centuries have been these:

1. Small mammals ate the dinosaur eggs and reduced the population.
2. A great plague destroyed the dinosaurs.
3. Starvation. Dinosaurs were big animals and required a lot of food.
4. Dinosaur bodies were too big for their small brains.

5. The climate changed from a tropical environment to one of extreme cold.
6. Volcanic eruptions blocked the sunlight, plants died, and the dinosaurs had little food.

Last, there's the asteroid theory, which is the most popular and widely accepted theory today. A large asteroid or comet struck the Earth about 65.5 million years ago. The father-son team of Luis and Walter Alvarez discovered a very distinct layer of iridium in rocks at a depth that corresponds to that time. Iridium is found in abundance only in meteorites and asteroids. Their initial discovery was in Italy in 1980. Since that time, that same kind of very thin layer of iridium has been found in various parts of the world at the same depth.

In 1991, geologists found the huge Chicxulub Crater (named after a nearby village) at the tip of Mexico's Yucatán Peninsula. The crater dated back to the time of dinosaur extinction. The meteor or asteroid crater was 110 miles across. Scientists estimate the asteroid was six miles in diameter and struck the Earth at 45,000 mph, with a force two million times the energy of the most powerful nuclear bomb ever detonated.

The heat from this asteroid or comet impact boiled the ocean's waters, ignited forest fires worldwide, and plunged the Earth into darkness as the debris blocked sunlight from reaching Earth. This dropped the Earth's temperature into the freezing range, killing most plants and animals. Plant-eating animals died a few months after the vegetation died. Only small animals that burrowed underground survived.

Even though the asteroid theory is the one held by most scientists today, a number of the other theories continue to find some support. There is evidence, for example, of a gigantic volcanic eruption occurring in India about 65.5 million years ago that sent a gaseous volcanic plume into the atmosphere. The reasoning is that this volcanic eruption was enough to trigger significant climate change, which the dinosaurs failed to adapt to.

Other scientists hold to the idea that there may have been more than one cause of dinosaur demise. They believe that both an asteroid impact and volcanic eruptions caused dinosaurs to die off. Or they hold to the possibility that an asteroid impact triggered volcanic eruptions and that the combination was deadly for dinosaurs, other animals, and plant life.

Dinosaur extinction was not the first massive die-off in history; the biggest was the Permian-Triassic extinction event, called the Great Dying, happened 251 million years ago. Over 70 percent of land vertebrates, and 96 percent of water species, were wiped out.

215. How did the Ice Age happen and when will be the next one?

Ice ages occur when the Earth has notably colder climate conditions. During an Ice Age, the polar regions are cold and there is a heightened difference in temperature between the equator and the north and south poles. The winters in the north regions are more extreme, allowing a lot of snow to accumulate, and the summers are cool enough that the snows of previous winters do not melt completely away. When this process continues for centuries, ice sheets begin to form.

The thinking right now is that the most likely cause of ice ages is the amount of radiation put out by the Sun. Variations in its intensity and timing are caused by changes in radiation of heat from the Sun; according to most scientists, these variations are responsible for the glacial cycles, and it may turn out that more than one cause responsible for ice ages. Ours is a very dynamic planet, and ice ages are most likely a result of complicated interactions among several of the below.

While the exact causes of ice ages are unknown and controversial, there is general agreement about *possible* causes. They are:

1. Changes in the amount of solar, or Sun, output
2. Changes in the amounts of carbon dioxide and methane put into the atmosphere
3. The motion of tectonic plates, meaning the shift in location of continents and oceans, which affect wind and ocean currents
4. Changes in the Earth's orbit around the Sun and the Sun's orbit around the center of the galaxy
5. The orbital motion of the Moon around the Earth
6. Impacts of large meteorites
7. Volcanic eruptions, including super volcanoes

Several of these factors can have reciprocal effects. For example, an increase in atmospheric carbon dioxide gas may alter the climate, and then the climate change may alter the composition of the gases in the atmosphere.

The last glaciers melted and receded from our area around Wisconsin about eleven thousand years ago. It's estimated that the next ice age will reach its peak in eighty thousand years, but nobody knows for sure when it will start. There have been ten periods of spreading ice sheets and falling seas in the last million years. Evidence comes from ice core samples from all over the globe.

Sometimes we forget about the time frame for planet Earth. All of human history is in our current warming period, called the Holocene. So it's hard to picture most of the United States and Europe buried under fifty feet of ice. Interglacial phases like the one we are experiencing right now are mere blips in between long periods of extreme cold.

These climate swings of big freezes interspersed with warming periods are rooted in changes in Earth's orbit and the tilt of its axis. Even though these changes are tiny, they have a huge impact on how much solar heat hits our planet. Right now we are in a warming period.

How does the knowledge of an upcoming ice age fit in with the current global warming? It might seem that human-caused global warming would fend off or delay the next ice age. But that may not be true, and it may even hasten one. Global warming could help bring on another ice age more quickly by melting the freshwater ice caps and shutting down the heat-carrying ocean currents that keep northern climates warm.

Most climatologists say that there is an extremely complex interplay of greenhouse gases, orbital shifts of the Earth's rotation, ocean currents, and output of the Sun, which is not well understood.

9

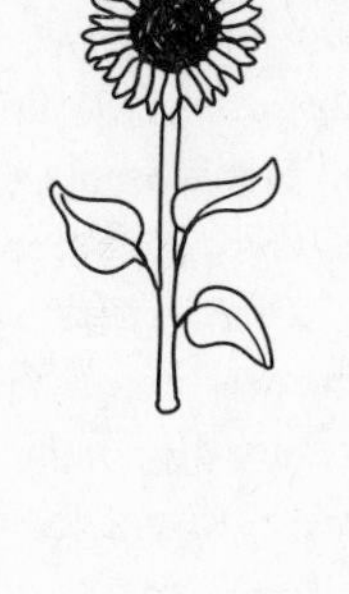

Plants, Animals, and Other Living Things

216. How does wood come from a seed?

It is a miracle, isn't it? A tiny acorn can grow into a mighty oak tree. One small seed has all the genetic code, directions, and information to become an exquisite tree with, its roots, trunk, and leaves. And water and nutrients can go from the roots in the ground to the highest branches of the tree.

Trees have flowers by which they reproduce. The egg in the pistil is fertilized by the sperm in the pollen of the stamen to produce a seed. The seed drops to the ground, and when conditions are right, the seed germinates, or sprouts, and the cells in the seed become active and start multiplying. The basic systems, such as roots, stems, and leaves, take form. The roots grow downward and the stem and leaves grow upward and break out of the ground. The tree is on its way!

We owe a great deal to trees. Trees allow us to live a comfortable life. Trees take in the carbon dioxide that we breathe out. And trees give off the oxygen that we breathe in. So trees do the opposite of what we humans do. At night, when photosynthesis can't take place, plants do use oxygen, but the amount is trivial. So the idea of removing plants from the bedroom because they take up a lot of oxygen is a myth.

Trees also help keep the amount of water in the air (relative humidity) at a proper level. They lend a hand in keeping the temperature about right, neither too hot nor too cold. The basic food that trees use to grow is carbohydrates, created in a complex process called photosynthesis, using sunlight and water.

Wood comes about by the growth of the cambium layer. New growth is added each year. A tree grows each new coat of wood over the old coat. The cambium layer lies just inside the bark and is sandwiched between the xylem and the phloem. These layers build up around the tree and can be seen in cross section when the tree is cut. We can count those rings to tell us the age of the tree.

The xylem carries water and nutrients from the roots up to the leaves. The leaves make the food, which is transported down the phloem to the roots. As growth continues, the xylem will become sapwood and the phloem will become bark. Tree sap is the general term for the fluid transported by the phloem carrying leaf-made food down to the roots. (There is some little sap in every family tree. A little joke here!)

Maple trees store starch and sugars in their sapwood. The rising temperatures of spring cause the pressure inside the tree to grow. When a hole is bored in the tree, sap drips out; this can be boiled down into maple syrup. One of the first signs of approaching spring, other than pitchers and catchers reporting for spring training, is the growth of the buds on maple trees. This bursting is noticed in February and is a response to the lower number of daylight hours.

There are over six hundred species of trees in the United States. Bristlecone pines can live for more than five thousand years. The redwoods of California can grow to a height of 380 feet.

Joyce Kilmer said it best: "I think that I shall never see a poem lovely as a tree."

217. Do fish sleep?

Yes, fish do sleep. Although it can be tough to tell, since they don't have eyelids that open and close, fish spend part of each day sleeping. You can tell when they're sleeping in fish tanks: They sit at the bottom and don't move, motionless except for minimal correcting motions with their fins to remain in position. They seem to be in a trancelike state of suspended animation. Scientists have recorded brain waves of fish. They show a distinct difference in the patterns between being awake and being asleep. To conduct your own test, you can drop food in the tank while your goldfish is sleeping; you will perhaps notice the fish takes longer to respond.

Fish sleep is a bit different than the sleep we humans enjoy. For most fish, it is a period of rest and reduced activity, not the deep rapid-eye-movement brainwave activity that occurs in humans.

Fish need the restorative nature of reduced activity and slower metabolism that comes with sleep. In that respect, they are the same as humans and most all others in the animal kingdom. Researchers kept some zebra fish awake by repeatedly giving them a mild electric shock. They found the fish suffered from sleep deprivation and insomnia. These pestered fish tried to catch up on their lost sleep as soon as they were left undisturbed.

Fish sleep behavior varies widely. Some fish wedge themselves in a spot in the coral or mud. Some build a little nest. The parrot fish secretes a mucus sleeping bag around itself. Other fish change color slightly, taking on a duller color so they are less noticeable. Some kinds of sharks have to keep moving because they need a steady flow of oxygen and water moving through their gills. The behavior of minnows drastically changes when they are trying to get some zzzzz's. They are very active in schools during the time they are awake, but they scatter and stay motionless during rest periods.

What do you call a fish with no eyes? Fsh.

218. Why do plants have roots?

Plant roots are like the heart and brains of a human: They perform the functions necessary to sustain life. Roots anchor the plant into the ground, provide and store the food the plant needs to grow and to reproduce, give plants the water they require to survive, and take in necessary minerals, such as nitrogen and sulfur. Plants can't move around the way animals do, so they can't go out and search for food. Instead, plants employ some marvelous and ingenious methods to endure and propagate.

Most plants have to make their own food. In a process called photosynthesis, plant leaves trap sunlight and use its energy to change carbon dioxide gas and water into a form of sugar called glucose. Glucose gives the plant energy and is used to make cellulose, which builds cell walls.

Plant leaves lose water by evaporation through tiny holes, or stomata. These microscopic openings also let carbon dioxide in and oxygen out. As the water is lost, more water is drawn up through the roots, in a process called transpiration.

You and I have a vascular system through which the heart pumps blood around our bodies. It is quite a complex plumbing system. Plants also have a vascular system, but it's much simpler, consisting of two types of microscopic tubes: the xylem transports water and minerals from the roots to the rest of the plant, and the phloem carries that glucose food to all parts of the plant. While most plants make their own food, there are rare exceptions, like the Venus fly trap. It grows in very poor soils and gets much of its food by ensnaring insects. The ends of the plant's leaves form a trap made of two pads with trigger hairs. When an insect lands on the pads, the jaws snap shut.

People have eaten roots for as long as there have been people. Native Americans ate camas in what is now the wheat fields in Idaho and Oregon. The camas plant looks somewhat like a sunflower, growing to one to five feet tall, but has lilac and blue-violet flowers. The camas bulb is dug in the fall and looks and tastes like a baked potato. Today, roots we eat include taro, yams, beets, rutabagas, and carrots.

The potato is an interesting specimen. All plants have three basic parts: root, stem, and leaf. The root is hidden under the ground. The stem is the woody part, usually tall, and it supports the plant. The leaves are green, flat, and thin, and they make food. But the potato is a stem in disguise—known as a stem tuber, it is a stem that grows underground and has some of the functions of a root. The potato has adapted over time to store food and water under the soil, where it is cooler than aboveground.

219. **Why do skunks smell so bad?**

Skunks have two glands, one on each side of the anus, that produce an oily secretion containing sulfur compounds known as thiols. The highly offensive smell is similar to a combination of burning rubber, rotten eggs, and strong garlic.

Skunks are reluctant to use their spray weapon. They will go through a routine of stomping their feet, hissing, displaying their black and white colors, and raising their tail high in the air as a warning. But once they decide to spray, up goes the tail, and muscles next to the scent glands can send an accurate spray a good ten feet. The average skunk carries enough chemicals for four or five uses. After that, it takes a week or more for the skunk to reload.

A baby skunk is called a kit. Kits are born blind and deaf, and their eyes open only after three weeks. They are weaned in two months but stay with their mother for one year. Because kits are so vulnerable, mother skunks are very protective of their young; this is when they are most likely to spray.

Skunks are nocturnal, which means they operate at night—and most often at dusk and early morning. Predators, such as wolves, badgers, and foxes, avoid skunks. One notable exception is the great horned owl; it has next to no sense of smell, so it will swoop down and grab baby skunks.

Skunks are omnivorous. They eat both plants and animals, such as insects, earthworms, rodents, snakes, birds, eggs, berries, roots, and nuts. Skunks will even go after beehives; their thick fur protects them from being stung. Since they'll eat anything, skunks like suburban areas, where they can get into people's garbage cans and bags. Skunks most frequently encounter dogs and cats in these rural and suburban tracts, so our pets are more likely to be victims of skunk spray than we are.

Skunks hear well and have a keen sense of smell, but they have terrible sight, which is why they get run over by cars and trucks. Those glands they carry are also flattened, causing them to release their odiferous and offensive chemicals.

So what is a person to do if they or their pet has been skunk sprayed? A time-honored technique was to bathe in tomato juice, but while tomato juice might help some, it only masks the skunk smell, which returns if you get wet.

220. Why do snakes bite?

Snakes bite for several reasons: to defend themselves when they feel threatened, when they are startled, or when they sense that they have no means of escape and to get food when they are hungry. Snakes will protect themselves, but most do not protect their territory or their eggs, and none protect their hatchlings. A group of snake eggs is called a clutch.

There are three thousand species of snakes worldwide, but less than 25 percent are poisonous. Australia has the highest percentage of venomous snakes.

Snakes do not bite or fight each other. However, some species, like king snakes and cobras, eat other kinds of snakes. Snakes rarely bite humans. If they do, their bite is seldom fatal. Half of all venomous snake bites are caused by handling or taunting or failing to stay away from them.

Legend has it that Cleopatra committed suicide by allowing an asp to bite her. It would more likely have been an Egyptian cobra. She supposedly did the deed upon hearing of the death of her lover, Marc Antony. Snakebite was a means of execution in the Middle Ages. The condemned was thrown into a snake pit and expired due to multiple bites.

The first antivenin, used to treat venomous snake bites, was developed in France in 1895. To make antivenin today, small amounts of snake venom are repeatedly injected into a horse. The venom initiates an immune system response, and serum containing the resulting antibodies is harvested from the horse's blood.

We had a lot of myths about poisonous snakes. One such tale was that a rattlesnake could strike you from a distance of about five times its length. Another bias we harbored was that snakes were slimy creatures. In fact, they are quite dry to the touch. How to treat a snakebite was also a part of our lore. Cut an "X" over the bite with your pocketknife, and suck out the poison. And do it quickly or you'll surely die!

The truth is that a rattlesnake bite is not a death sentence. Staying as calm and still as possible, keeping the wounded part lower than the heart, abstaining from drinking or eating anything, and getting to a hospital are recommended.

The Scheckel boys were no fans of snakes. Occasionally a snake would come up the hay loader and land in the hay wagon. Other times the grain binder would package a scaly critter into a grain bundle. It was our duty to rid the farm of these vile varmints!

And our thinking was partially Bible based. After all, was it not a snake that tricked Eve into eating the forbidden fruit? A person's thinking evolves over time. Even though I prefer dogs over snakes, snakes probably do have their place in God's Kingdom.

221. How do flowers get their colors?

The pigment anthocyanin gives color to most flowers and fruits. Anthocyanins are water-soluble pigments in a class of chemicals called flavonoids. A pigment is an organic compound that gives a characteristic color to plant or animal tissue. For example, chlorophyll gives the green color to plant leaves and stems. Hemoglobin gives blood its red color.

Flavonoids, compounds found in many vegetables and fruits, have antioxidant properties. Some flavonoids protect blood vessel walls, some alleviate allergies, and some defend against cancer and viruses. Others have anti-inflammatory properties.

The flower colors of blue, purple, pink, and red come from anthocyanins. Plants produce other pigments, too, like carotene, which makes the orange of carrot roots and the red of tomatoes; chlorophyll, which gives leaves their green color; and xanthophyll, which makes foods like egg yolks and corn yellow.

A common experiment uses the anthocyanins in red cabbage as a pH indicator, because anthocyanins change color depending on their pH. A strip of paper treated with red cabbage juice will turn red if placed in an acidic (low-pH) solution and green/yellow if placed in a basic (high-pH) solution. The strip will remain purple if placed in a neutral-pH solution.

Colors are instrumental in a flowering plant's reproduction. Flowers reproduce by having male organs called stamens, loaded with pollen, and female organs named stigmas. The function of the stigma is to receive pollen from another plant of the same species. But plants can't move, so they rely on insects and birds to move their pollen for them. The bright flower petals attract insects and birds to the nectar or edible pollen produced by the plants. Yellow flowers are an advertisement for bees, and red attracts birds.

Some flowers change color during the growing season. For example, forget-me-nots change from pink to blue. Larkspur, or delphiniums, change color, too. Color shifts send a signal to insects that a flower has aged and is past pollination.

Certain flowers close up at night. The morning glories that had to be pulled out of the cornfields on my farm closed up at night and opened during the day. These flowers react to changes in light and temperature. Heat makes the inner surface of the flower grow faster, and the flower opens. When the temperature goes down in the evening, the outer surfaces grow faster than the inner and the flower closes up. Flowers that respond to light have cells that contain a pigment called phytochrome. These photoreceptor cells increase and decrease in size based on the amount of incident light.

The main reason people grow flowers is for their colors and the feelings these colors evoke for weddings, funerals, holidays, gifts, birthdays, anniversaries, and dates. We think of lilies for Easter, poinsettias for Christmas, roses for Mother's Day. The red of the rose generally stands for love, pink for gentility, happiness, grace, and thankfulness. Daffodil or chrysanthemum yellow signifies joy and friendship. White is for purity, innocence, silence, and heaven. Coral- or peach-colored flowers are common for a first date. Purple flowers demonstrate royalty and ceremony. There are claims that the blue of hydrangeas and irises calms worries and represents peace and serenity. However, the science behind these claims is dubious.

222. Do cats see in black and white?

It was once thought that cats were color-blind, but scientists now know better. Tests show that cats can tell the difference between certain colors. Cats can see some colors, and can tell the difference between red, blue, and yellow light, as well as between red and green light. They have difficulty distinguishing between colors on the red end of the visible spectrum.

Research indicates that cats can distinguish more grades of gray than humans. They see less saturation in colors, so they don't experience colors as vibrantly or intensely as we humans. But cats don't pay much attention to color. In their history and ancestry, color was not important to their survival. Motion was everything. So cats see even the slightest twitch.

Can cats see in the dark? Not in total darkness. But cats are nocturnal hunters, and they are able to use even the smallest glimmer of light.

They see much better than humans in very dim light or semidarkness, although their vision is blurred at the edges, and they see best at a distance of between six and twenty feet.

Cats have a membrane, called the tapetum lucidum, that lies behind the retina. This layer of specialized cells reflects light back through the retina, thus increasing the light available to the photoreceptors. Photoreceptors are cells, such as rod cells, that change light to electrical impulses (see page 48). This improves vision in low light conditions. Many night hunters in the animal kingdom are endowed with a tapetum lucidum. In cats, the tapetum lucidum improves vision by a factor of six. Cats can see things in the dark that are invisible to humans. Reflection of light from the tapetum lucidum explains the characteristic green or gold glow from a cat's pupil.

We had a cat down on the farm that was particularly adept at hunting gophers. That calico barnyard cat would crouch motionlessly for hours just a few feet from a gopher hole and then pounce on the hapless striped rodent, shake it to death, and then bring it around the buildings, where we would spot that cat with a gopher in its mouth.

We kids liked that gopher-killing machine because there was a bounty on gophers, moles, and rattlesnakes. The cat would drop the gopher at our feet, and we would cut the tail off the gopher and preserve it by putting it in a canning jar generously sprinkled with salt. The jar of gopher tails was stored in the garage. Once the tail was removed, we would return the gopher to our moneymaking cat. We figured if the cat didn't get the gopher back, it would simply stay away from the farm buildings and eat the gophers out in the pastures.

Gopher tails fetched a nickel and moles' feet (just the front ones, and you had to have both) brought a quarter. The town clerk was authorized to pay the bounty. Unlike the other farm cats that guzzled milk, bread, and leftovers, this cat earned its keep. This was no welfare cat!

223. Humans can't grow back arms or legs, so why can some animals, like starfish, regenerate their limbs?

Starfish have incredible powers of regeneration. If you chop off an arm, it will grow back within months. If you chop a starfish in half, it will grow back into two starfish. Some starfish can regrow a new central disc from a single arm. For some species, a portion of the central disk must be present to regrow a whole new starfish.

The "chopped-off" part of a starfish begins by growing tiny budding arms from the severed piece. The first development is the forming of a crescent-shaped ridge at the end of the severed part. Grooves begin to form, and a mouth develops at the point they radiate from. The arms begin to grow, and tube feet start to appear. A separated limb lives off stored nutrients until it grows a disc and mouth and is able to feed again. Complete growth can take a year or more.

Starfish are echinoderms, a family of seven thousand spiny marine creatures. Sea urchins and sea cucumbers belong to the same family. There are at least fifteen hundred different types of starfish. Marine biologists call them "sea stars" because they are not actually fish.

Most starfish have five arms, but some have many more. The sunflower starfish, found off the coast of Britain, has twenty-four arms and can get up to more than a yard across. But the most common ones, the kind we find dried up in gift shops, are about the size of your hand and have the familiar five arms.

Starfish are a marvel of engineering. They have a network of tubes that carry seawater from a hole on their top side to hundreds of tiny feet, which lie in rows along each arm. Each foot is a hollow tube filled with sea water. The vascular system of the starfish is also filled with sea water. By moving water from the vascular system into the tiny feet, the starfish can make a foot move by expanding it. Muscles within the feet are used to retract them. They pump water in and out of these tubes, so a starfish can move each foot or arm independently, propelling itself along the seabed.

The tip of the arm has a primitive eye, called an eyespot, that allows it to see light and dark, so it can detect movement.

They are also very strange eaters. Their mouth is on the underside of their body. They actually turn their stomach inside out, a process termed "everting." They clamp their tube feet around a mussel and pull the shell apart. Then they evert their stomach, much like turning a pocket inside out. They push their stomach into the mussel shell to suck all the juices out of it. It's an ingenious way of not having to deal with the hard, solid shell material.

224. How are bumblebees able to fly with such small wings and a big body?

Yes, the minuscule wings on the gargantuan body of the bumblebee have intrigued biologists and flight aerodynamics people for decades. But modern science has some remarkable tools to probe the secrets of "The Flight of the Bumblebee." They have employed high-speed digital photography and robotic models of bee wings to come to some basic understanding of how these remarkable creatures can not only fly but also hover. Most amazing of all, bumblebees can carry their own body weight in pollen and nectar back to their hive.

The bee takes in oxygen through twenty-four breathing holes and uses the same amount of oxygen, for its body weight, as a bat or a bird.

Bumblebees have a very rapid wing beat—about 200 times every second, well over eight hundred thousand times in an hour. That's faster than a politician flaps his jaw! The bee's wings flap rapidly from nerve impulses. The vibration is much like a stretched rubber band that is plucked. Each nerve impulse causes ten to thirty movements of the wing. The lift created by a bumblebee's wing beating is a bit different from the lift created by a bird's wing flapping or an airplane's flight. In the bee, the air swirls over the edge of the wing and creates a small eddy—basically, a low-pressure vortex, which sucks the wing upward and generates lift.

What do bees do differently when they take on a load of nectar and pollen? They do not make their wings beat faster. Instead, they stretch out their wing surfaces slightly, creating a bit more vortex, which increases lift. Scientists figured out this bee secret by putting bees in a chamber of oxygen and helium. The oxygen allowed the bees to breathe, but the less dense helium made them work harder.

There is the possibility that unlocking the secrets of bee flight can lead to designing aircraft that can hover in place for military surveillance, monitoring disasters such as earthquakes and tsunamis, or carrying loads of relief supplies.

225. Why do birds chirp or sing?

The predominant reason birds sing at dawn is to stake out their territory. They are saying, "I made it through the night, and I'm still here." At dawn, the air is generally calmer and there are fewer distracting sounds. Sound carries better in the cool, dense, still air, too. It has been shown that a bird's song in the wee morning hours is twenty times more effective in keeping other birds away than the same song at noon. One hour before sunrise is prime time for bird singing.

The singing starts at different times for different species of birds. We've all heard robins sing at three o'clock or four o'clock in the morning. Biologists believe that the robin senses that it has a better chance of being heard when there is little competition around. Nightingales and wrens follow the same pattern. Most bird singing in the waking hours is done by males. That does not hold true for all bird species. Female robins, female Baltimore Orioles, and female rose-breasted grosbeaks can belt out a tune with the best of the males. It is the same for many female owls.

Chirping isn't always territorial; bird singing is louder and more frequent in the mating season. When the mating season comes to a close, morning singing dies off a bit. Much time is needed for housekeeping, so there is less time for singing.

An additional school of thought supposes that the low light level in the predawn hours makes them a bad time for foraging for food, so perhaps, with nothing better to do, birds find it's a good time to sing. Some birds prefer to warble away at dusk rather than dawn. Sparrows, for example, are very vocal in the early-evening hours.

We humans learn to talk by listening to other humans talk. It is nearly impossible for someone born deaf to know what most sounds should sound like. It is somewhat different for birds. Young birds do sing from the time they hatch, but they refine their songs by listening to adult birds. It requires several months for the young birds to develop a tune.

Birds have a special sound-producing organ called a syrinx. It is equivalent to the human voice box, or larynx. Air from the bird's lungs passes over the muscles and membranes in the syrinx. The membranes vibrate and produce the sound.

A few birds sing when flying, but they are in the minority. While most of us can walk and chew gum at the same time, most birds have to alight to sing.

The starling is an amazing species. Starlings will mimic sounds they have heard, including car horns, police sirens, and telephone ringtones. They will even work pieces of those songs into their singing.

I'll end with a very bad bird joke: What do you get when you run over a robin with your lawnmower? Shredded Tweet.

226. Why do sunflowers always face the Sun?

Sunflowers exhibit heliotropism. "Helios" is another word for sun. "Tropism" means a turning movement in response to an environmental stimulus. Heliotropism is the daily motion of plant parts in response to the movement of the Sun. Heliotropic sunflowers track the Sun across the sky from east to west. At night, the sunflower head is oriented in a random direction. But at dawn, it turns again toward the east. Sunflowers show this Sun-following tendency only in their bud stage. In the blooming stage, heliotropism ends and the stem is frozen, typically toward an eastern direction.

The movement of the sunflower head is performed by a group of motor cells in a flexible segment just below the flower head, called the pulvinus. The motor cells force potassium ions into tissues, thereby increasing their cell pressure. The segment can flex because the cells on the side in shadow elongate due to this rise in pressure.

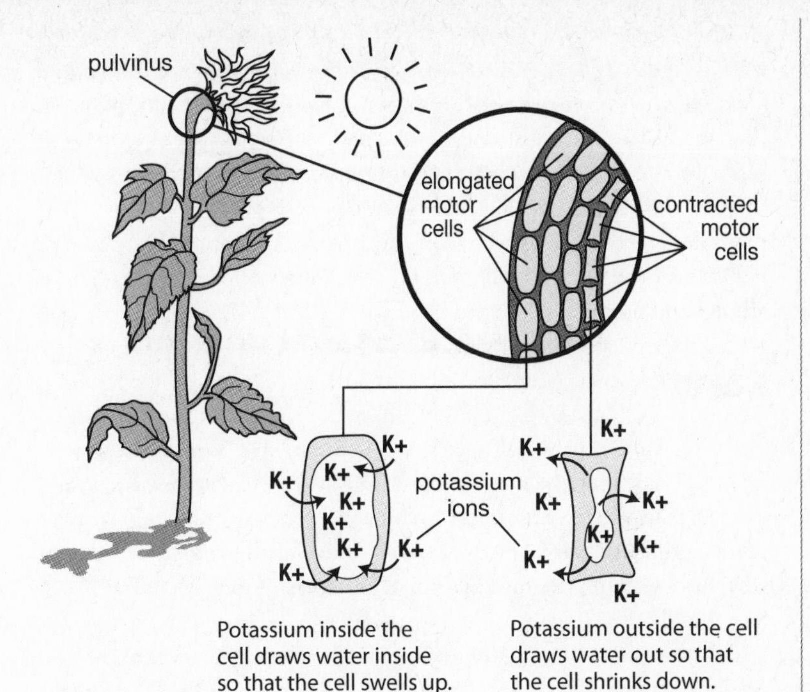

The visible part of sunlight is actually made up of seven colors: red, orange, yellow, green, blue, indigo, and violet. Heliotropism is a response to just the blue light contained in sunlight. If the sunflower is covered with a red transparent sheet that blocks the blue light, the sunflower head cannot follow the Sun.

In the 1530s, Francisco Pizarro found that the Inca Indians of Peru venerated the sunflower as an image of their sun god. Today, sunflower seeds are sold as snacks and bird food. The seeds are sometimes processed into a peanut butter alternative, which is safe to eat for people who are allergic to peanuts. Sunflower oil is used for cooking and has been blended into biodiesel fuel. After the seeds are removed, the sunflower head can be used for livestock feed. The tubers of one species are eaten as a vegetable; these are known as Jerusalem artichokes. The state flower of Kansas is the sunflower. An Alpine plant known as the snow buttercup is also heliotropic.

In Greek mythology, the water nymph Clytie fell in love with the sun god, Apollo, also known as Helios. Apollo, having once loved her, abandoned Clytie for another. She sat on a rock for nine days, without food or drink, staring at the Sun and mourning her loss. She watched his chariot move across the sky. After the nine days, she was transformed into a sunflower.

A few years ago, while traveling through North Dakota, we saw huge fields of sunflowers. It was a beautiful sight!

227. What were the first dogs on Earth?

The dog belongs to the same family as wolves, coyotes, foxes, and jackals. The family name is canids, or canines. All canines share these characteristics: They bear young by live birth, have similar dental structure, walk on the toes instead of the soles of their feet, and maintain their body temperature at a constant level.

The tamed, or domesticated, dog has been around for almost all of recorded history. It was the first animal to be a companion to humans. Dog bones discovered in the United States have been dated to 8300 BC.

Scientists say that the dog's natural hunting instincts have been used by humans in varying environments and that this led to the development of different dog breeds. There is evidence that the first breeds of dogs were "sight hounds" or "gaze hounds" that had deep chests, long legs, and a keen sense of sight. These dogs could spot their prey at a far distance and then sprint quickly and silently to run it down in open, treeless country.

The oldest breeds of dogs are African and Asian and include the basenji, Lhasa apso, and the Siberian husky. "Scent dogs" have a large nose with well-opened nostrils. In Europe, they were bred for stamina and used to track prey over a long period of time. The British hunting foxhound is a prime example. Breeders in Britain developed dachshund dogs to hunt burrowing animals such as badgers, foxes, and rabbits and to control many types of vermin. These dogs, known as terriers, needed to be feisty and energetic.

Later, when guns were invented, dogs with sensitive noses were bred to locate and indicate the location of prey, flush out the prey, and retrieve it once it was shot. These dogs had to have a soft mouth. They were called pointers and retrievers.

One of the most loyal dogs is said to be the golden retriever. According to the American Kennel Association (AKA), the Labrador retriever is "top dog" in numbers of registrations and the German shepherd is number two.

When I was a kid growing up on the farm, I had a brown dog named Browser. He was a mutt, or crossbreed, but I didn't care. He would chase the cows and bring them in for milking. Browser was a good squirrel-hunting dog. I could talk to Browser and tell him things and he wouldn't blab it around.

I would say the most lovable dog is the basset hound. My wife and I have had two of them, Mux and Tater, in the years we've been in Tomah, Wisconsin. They're both in dog heaven. That's what we told our grandchildren.

I like the saying or prayer that goes something like this: "Please, God, help me be the person my dog thinks I am."

228. Where can you find bacteria?

Bacteria are all around us: in the air, in water, in food, on our skin, and inside our bodies. They are so small they can only be seen by using a high-powered microscope. There are no male and female bacteria; they multiply by splitting. And if they have the right conditions of food and temperature, they keep right on multiplying.

Bacteria have only one cell. The outside is a membrane, like a skin, and the inside is called cytoplasm, usually without the central core, or nucleus, that plant and animal cells have. In some bacteria the membrane is slimy. Many kinds of bacteria have tiny tails, called flagella, that allow the bacteria to move. Others move by extending and shrinking their cell bodies to glide or twitch, much the way worms move. The skin, or membrane, is not waterproof. The only way to destroy bacteria is for chemicals to get through that skin and attack them.

Bacteria can infect us and cause diseases, but bacteria can also be very useful. Long before people knew what bacteria were, they were familiar with their effects, such as fermenting wine, making milk sour, and making dead plants and animals decay.

In order for bacteria to do bad things to us, they have to get into our system. Our bodies have a whole army of defenses. Dry skin, stomach acids, hand washing, and tooth brushing all help to make life difficult for unfriendly bacteria. They can enter the body through damaged tissue, wounds, bites, or the mucus barrier that lines the nose and the alimentary canal (mouth to anus). Once such bacteria get into our system, they are free to grow and spread. Usually, they start as small local infections and spread rapidly once in the bloodstream. Most bacteria inside us never make us sick. For example, we all have some *E. coli* bacteria in our intestines. But sometimes the *E. coli* can acquire a gene that causes them to secrete a toxin; it's in these cases that we get sick (see page 288).

Toxins that are secreted by or leak from the bacterial cell into our bodies are named exotoxins. Diphtheria, cholera, and tetanus are diseases resulting from different types of exotoxins. Other diseases caused by bacteria include leprosy, typhoid fever, plague, tuberculosis, anthrax, Lyme disease, tooth decay, and tonsillitis.

Antoni van Leeuwenhoek was the first person to see bacteria, in 1674. He had started grinding glass lenses to make better magnifying glasses so he could examine the cloth at the draper's shop where he worked. Leeuwenhoek turned his magnifying glasses on pond water and saw little critters (ptotozoa) in it. He made detailed drawings of his many discoveries.

229. Which animals are the most intelligent?

The chimpanzee would be tops on every researcher's list. They can make and use tools, band together to hunt, solve advanced problems, learn sign language, and use symbols for objects. They remember the signs for the names of people they have not seen in years, form strong bonds with members of their own kind, and adhere to a strict social order. They engage in acts of violence, but also show empathy to other chimps—kind of like humans!

The orangutan is in the same "great ape" category as chimpanzees. They use tools and have a very strong culture. The young stay with their mother for many years, and she teaches them all they need to know to survive in the forest.

Gorillas will grab a tree branch and use it to gauge the water's depth before crossing a stream or bog. They have been observed utilizing a small log as a bridge to walk across a muddy patch. Apes will employ a rock to crack open a nut.

Beyond the apes, there is no agreement on smartest animal, but five or six would certainly make the top ten. Dolphins are very sociable, have a complex language, and can learn a vast array of commands to perform a wide variety of tasks. That is why they are a favorite at aquariums. They also seem to have more fun than many creatures with their leaping, whistling, racing, spinning, and surfing.

Pigs have gotten a bad rap for their seeming lack of hygiene and gross gluttony, but they are very smart animals and as trainable as cats or dogs. They can adapt to a large variety of ecological conditions and hence are found all over the world. Christopher Columbus brought the first pigs to the New World in 1493 on the second of his four voyages. They adapted and multiplied rapidly. Pigs are also, in fact, among the cleanest animals around. They go to specific areas away from where they sleep and eat to urinate and defecate. They have no sweat glands, so they wallow in the mud to stay cool. Pigs can move a cursor on a video screen with their snout and pick out objects on the screen.

Crows and jays are members of the corvid family, and clever creatures they are. They communicate in elaborate dialects and play games and tricks on each other. Crows adapt to almost any condition and live just as well in the countryside as in large cities. They will take nuts from trees, place them in the street for cars to run over them to crack the shells, wait for the light to change, and then swoop in to get the soft nut meat inside.

Elephants clean their food, are curious, and follow human commands in captivity. Showing empathy is considered an advanced form of intelligence, and they have been seen consoling family members. Some elephants recognize themselves in a mirror.

It's hard to believe that squirrels have a high intelligence, especially when we see a squirrel dart across the road in front of our car and get most of the way across the road, only to double back and get flattened.

But squirrels are cunning furballs, with great persistence and incredible memories. They will steal food from bird feeders, store and cache food for lean times, and find their hidden treasures—about ten thousand of them, all buried separately—months later. Those clever devils will even pretend to hide food to confuse potential thieves.

230. Do pet owners and their pets understand each other?

About two thirds of dog owners think their pet understands them, which is higher than the percentage of parents who understand their teens! Just a little joke there! This high level of communication shows a strong bond between owners and their pets, although there seems to be a particularly strong bond with dogs—only about 50 percent of cat owners report the same kind of connection.

Experts in animal behavior say that animals and people learn to communicate by sound over time by associating certain sounds with actions, such as a particular bark when a dog wants to go outside. Animals respond to certain words, such as "get off the bed," "roll over," "stay," "fetch," or "come here."

What almost all pets do understand is tone of voice and, to some extent, facial expression and body language. Of course, the same is true of people. Tone of voice conveys a lot of meaning for both pets and humans.

According to a local veterinarian, pets are experts at reading body language. Pets are able to read moods, such as happy, sad, or angry, all from the movements of our bodies. Dogs seem to be better at nonverbal communication than cats. And some breeds are better at it than others. Border collies, poodles, German shepherds, and golden retrievers get high marks for reading their owners' demeanors. The vet I spoke with recommends two books for further reading on the subject: *The Other End of the Leash*, by Patricia McConnell, and *How to Teach a New Dog Old Tricks*, by Ian Dunbar.

Animals are not equipped for speech like we humans are. They have tongues, but not the complex vocal cords that we humans have in our throats. So they can't make smooth vowel sounds. Animals lack the speech center in the brain that we humans possess. Parrots approach the ability to speak like us, but they are mimicking what we say.

Lots of animals "talk" to each other. Birds do this when they chirp and sing, dogs bark, cats meow, and dolphins click and chatter as they play. Bees do a complicated dance that tells other bees which direction to go for food. Gorillas thump trees to lead others through a forest.

231. Why does a dog wag its tail when it is happy?

For dogs, tail wagging serves the same function as a human smile or a nod of recognition. A dog will wag its tail for a person or for another dog, or even a cat, a mouse, or a horse. A dog seems not to wag its tail for any lifeless thing. Tail wagging, then, is a form of communication.

According to Dr. Stanley Coren, author of the book *How to Speak Dog*, a dog's tail was designed to assist in its balance. When a dog makes a quick turn, it throws its tail in the same direction that the body is turning, thus preventing the dog from spinning off course. Dogs also use their tails to walk along narrow surfaces. They swing the tail in the opposite direction of any tilt of the body to help maintain balance. A tightrope walker does the same thing by using a balance bar.

Puppies do not wag their tails until they are about thirty days old. Before that time, just about all they do is eat and sleep, and there is no need for them to communicate. At about six or seven weeks, the puppies are socially interacting with one another. They play, fight, nip, jostle, chase, and cuddle. They learn to use their tail to signal their intentions and to prevent conflicts. Tail wagging often acts as a truce flag. Later, puppies learn to wag their tails to beg for food. Tail wagging is like a smile in another way: When a dog wags its tail in that familiar, broad-stroked, side-to-side manner, it means the dog is happy and content.

Sometimes a puppy will chase its own tail. The puppy is just playing by itself, since there may not be any human or another puppy to play with. The puppy probably doesn't yet realize that this tail is its own. Usually, this is something a dog will grow out of.

There is also the theory that tail wagging, unlike barking, is a form of silent communication. When early dogs were hunting, a bark would scare the game away, but a tail wag served as a silent signal to other dogs that a meal was close at hand. Dogs probably didn't think this idea through; it was more likely a matter of instinct.

A dog joke thrown in for good measure: Why did the poor dog chase his own tail? Because he was trying to make ends meet!

232. How do scientists know how old dinosaurs are?

Dinosaurs hold a fascination all their own; they're the mightiest land animals ever known. Dinosaurs were reptiles that lived in the Mesozoic era, millions of years before there were humans. Most hatched from eggs, none lived solely in water, and none could fly. The smallest was the size of a chicken; the largest were one hundred feet long and fifty feet tall.

Some walked on two legs (bipedal), some on four (quadrupedal). Some were armor-plated, and some had horns, crests, spikes, or frills. Some ate meat, some ate plants, and some did both. The dinosaurs dominated the Earth for over 160 million years. But then it all went south!

Dinosaurs suddenly went extinct 65.5 million years ago (see page 293). It was the end of the Cretaceous period, a time of high volcanic and tectonic plate activity. There are several theories, but the most widely accepted is that an asteroid impact caused major climatic changes. The dinosaurs couldn't adapt.

Evidence of dinosaur extinction 65.5 million years ago comes from the discovery in 1980 of a thin layer of iridium-rich clay. The iridium was attributed to an asteroid or comet impact near what is now Yucatán Peninsula (see page 294). Dinosaur fossils are found below the iridium layer, but not above the layer.

Dinosaur fossils are found all over the world. No single technique is used to date dinosaur bones. I am always amazed at these huge creatures that roamed the Earth long before man became a dominant force on the planet. Dating dinosaur fossils is fairly straightforward. The oldest technique is stratigraphy, or studying how deep the fossil is buried. Rock layers, or strata, are deposited horizontally over time, and new layers are formed over older layers. Based on knowledge of the history of these layers, scientists estimate the amount of time that has passed since the layer containing the fossil was formed. Observing the changes in the Earth's magnetic field (magnetostratigraphy) is another technique. Different magnetic fields in rocks are from different geological eras. Radioisotope dating of igneous rock found near the fossil is a third procedure. Uranium-235 decays (changes) to lead-207 over a known constant rate. Scientists get a good estimate of the age of the rock by examining the ratio of uranium to thorium or the ratio of potassium to argon. Looking for so-called index fossils is a fourth practice for dating dinosaur bones. Some common fossils are widely distributed but are quite time specific. For example, brachiopods appeared in the Cambrian period and ammonites are from the Triassic and Jurassic periods.

The first bones recognized as dinosaur bones were found in England in the early 1800s. The term "dinosaur" comes from the Latin words "*deinos*," which means "terrifying," and "*sauros*," which means "lizard." *Tyrannosaurus rex* had a very large head, with great jaws filled with double-serrated teeth. Its hips and hind limbs were massive, so all *T. rex* had to do was grasp its prey in its mouth and tear it apart. Terrifying, indeed!

233. How come parrots can talk but other animals can't?

Parrots (also mynahs, crows, and ravens) are famous for their ability to imitate different sounds that they hear. Most biologists and scientists think that parrots, along with ravens, crows, jays and magpies are among the most intelligent birds.

Unlike us humans, parrots don't have vocal cords. They learn to control the movement of muscles in the throat so as to reproduce certain tones and sounds. Air is expelled across the mouth of the bifurcated (two separated branches) trachea. Different sounds are produced by changing the depth and shape of the trachea. Some biologists speculate that parrots can "talk" because of the structure of their tongues, which are thick and large. However, mynah birds can mimic human sounds, too, and they don't have large, thick tongues. Other people theorize that parrots' voice mechanisms and hearing work more slowly than other birds' and that the sounds humans make closely resemble the sounds parrots naturally make.

Indeed, even though birds can be trained to do certain things, they can't reason or have real humanlike conversations. Parrots do not know the meaning of the words they repeat. Calls and songs are innate in birds; there is no learning involved. A bird raised in complete isolation and without ever hearing another bird will produce its own "song," which will be characteristic of that species.

Parrots are remarkable creatures. Some are barely three inches tall, and others grow to over three feet in height. They show colors of green, yellow, blue, and red. Parrots can adapt themselves to practically any living conditions, which is probably the reason sailors have taken parrots to sea for centuries. They are typically tropical birds but can adjust to temperate and even cold climates.

234. Why do cows rechew their food?

Although I grew up on a Wisconsin farm, cows are a little out of my field (no pun intended). So I turned to a local agriculture instructor to make sure my facts are straight. Cows' stomachs have four compartments, but not four separate stomachs. They graze on grass and swallow it half-chewed, and then they store it in the first compartment, called the rumen.

In the rumen, grass or hay is softened by fluids and formed into small wads, or cuds, before it is regurgitated to the mouth, where the cow chews it forty to sixty times, for nearly one minute, before the cud is returned to the rumen. Then it moves into the second compartment, or reticulum. That rumen word is interesting. Perhaps that is the origin of "ruminate," to chew the cud, or to think deeply about something or meditate, reflect, muse, ponder, or contemplate.

From the reticulum, the cud progresses to the omasum, where it is pressed, broken down further, and filtered before moving to the fourth stomach, called the abomasum, where it is digested. Finally, the grass goes into the intestine, where the cow takes out all the nutrients she needs to produce milk and keep her healthy and contented. It takes seventy hours to turn grass into milk.

If the cow eats corn (more easily digested), it is a straight shot from rumen to reticulum to omasum and finally to abomasum. From an evolutionary standpoint, cows developed four stomach compartments so that they could graze quickly and then go hide from their enemies, where they could then take plenty of time to digest their greens.

235. Why do dogs drool?

Since I'm not exactly an authority on dog drool, I consulted a local veterinarian to make sure I got this right. Some dogs are natural droolers. The basset hound, bullmastiff, and Saint Bernard are classic examples of dog breeds that regularly drool and slobber. These dogs have big, heavy lips, and the skin around their mouths and jaws is very loose and floppy, which lets the drool seep out, especially when they eat and exercise. Evaporation of saliva aids in cooling the big canine.

But sudden drooling in a pet that doesn't usually drool or slobber indicates a problem. The problem can range from a chipped or cracked tooth or a gum infection (the more common reasons) to poisoning or a foreign object lodged in the throat. Try checking your dog's mouth and looking to see if you can spot the problem. Maybe there is a splinter or foreign object you can see and easily remove without hurting the animal. If not, then it's time to visit the vet and have them diagnose the severity of the condition.

Dogs may drool because they are panting. Panting is one way a dog stays cool and gets rid of excessive body heat. Dogs can't sweat like we do. They wear that heavy thick fur coat. Dogs sweat through their paw pads, and it's by panting that dogs circulate the necessary air through their bodies to cool down.

There have been some great "dog-drooling" movies over the past years. *Beethoven*, with Charles Grodin and Bonnie Hunt, came out in 1992. They're the parents of three kids who adopt a big, slobbering Saint Bernard. The evil vet was played by Dean Jones, who starred in a lot of old Disney films.

I won't soon forget the 1989 film *Turner and Hooch*, with Tom Hanks and Craig T. Nelson. Hanks plays Detective Scott Turner, who must take care of Hooch, a slobbery Dogue de Bordeaux, or French mastiff—a relative of the bullmastiff and the Alpine mastiff—which witnessed a drug-related murder.

I consider *Homeward Bound: The Incredible Journey* the best "animal movie" ever made. This 1993 movie is based on the 1963 Sheila Burnford animal adventure novel *The Incredible Journey*. Shadow, the golden retriever (voiced by Don Ameche); Chance, the bulldog (voiced by Michael J. Fox); and Sassy, the Himalayan cat (voiced by Sally Field); struggle to find their way home. You won't see more beautiful wildlife and location scenery. Do you recall Sassy's famous line: "Cats rule and dogs drool"?

10

Sound and Music

236. What is a sonic boom?

Toss a pebble into a pond of water and little waves will form concentric circles that move away from the point of impact. If a boat travels through the water, waves will propagate both ahead of and behind the boat, and the boat will travel through those waves.

But if the boat goes faster than the wave travels, the waves can't get out of the way of the boat fast enough, and they form a wake. That wake is a larger single wave, formed from all the little waves that would have propagated ahead of the boat if the boat weren't going so fast.

Sound travels at about 750 mph. At the moment when an airplane moves faster than sound, it produces a sonic boom. That boom is the wake of the plane's sound waves. All the sound waves that normally would travel ahead of the plane are combined together, because the plane is traveling faster than the sound it's making, so the waves are forced together, or compressed. So at first you hear nothing, then you hear the boom that the waves create.

Sonic booms from aircraft generate an enormous amount of energy and sound much like an explosion. The loudness of the sound depends largely on the distance between the listener and the aircraft. It is usually heard as a deep double boom. A shock wave comes off the nose and leading edges of the plane, and another emanates from the tail. People in Florida heard that double boom every time the space shuttle returned from Earth orbit and landed at the Kennedy Space Center.

The crack of a supersonic bullet passing overhead is an example of a miniature sonic boom. So is the crack of a bullwhip. The whip is tapered down from handle to tip. The velocity of each part of the whip increases with the decrease in mass, so that the tip travels faster than sound travels.

We say that the speed of sound is 750 mph. But the speed of sound depends on the temperature, humidity, and pressure of the air. The speed of sound at thirty thousand feet is around 670 mph.

The speed of many aircraft is given as the Mach number, where Mach 1 is the speed of sound. So a plane traveling at Mach 1 is going about 750 mph, Mach 2 about 1,500 mph, and Mach 3 around 2,250 mph. The only passenger plane that was supersonic was the British-French *Concorde*, which flew from 1976 to 2003. It flew at Mach 2 and made flights between London, Paris, Washington, New York, and other cities.

Chuck Yeager was the first to break the sound barrier, in the Bell X-1, on October 14, 1947. The bright orange Bell aircraft, shaped like a .50-caliber bullet, is on display at the National Air and Space Museum in Washington, DC. It hangs over the 1903 Wright Brothers *Flyer I*, close to Lindbergh's *Spirit of St. Louis*, and above *Columbia*, the Apollo spacecraft that took men to the moon in 1969.

237. Is music mathematically based?

Math and music share many similarities: They are creations of the brain. Neither is tangible. You can't see them or touch them, the way we can wood, dirt, or cotton. Both math and music are abstract constructions taking place in the brain.

And, yes, music is based on math, too. There are two earmarks in music. One is the note known as concert A, the note orchestras tune to, which, when played at the octave near the center of a piano's keys, is 440 Hertz (Hz), sometimes referred to as 440 cycles per second (cps). On a piano, it is five white keys to the right of middle C. Middle C is 262 Hz. We can generally say that pitch, tone, and frequency are all the same thing, referring to the number of vibrations per second, and the particular rate of vibration of any given frequency creates a particular series of sound waves that we hear as recognizable musical notes.

The other important math feature of music is the twelfth root of two, which is 1.059453. Using 1.06 is close enough. Each note is separated by the number 1.06. When we say the twelfth root of two, we're asking this question: What number multiplied by itself twelve times gives two? And that number is 1.06.

If we take the note with a frequency of 440 Hz and multiply it by 1.06, we get 466 Hz, which is A (A sharp). If we multiply 466 by 1.06, we get the next higher note of B, which is 494 Hz. Multiply 494 by 1.06, and we get the next note, C, at 523 Hz.

There are twelve notes in an octave, each a half step, or semitone, apart. The piano is a very acoustically rich instrument. Its eighty-eight keys cover a tad over seven octaves. The white keys are the natural notes, and the black keys are the sharps and flats. In this system, a sharp is a pitch a half step higher. A note of F sharp is one half step higher than an F note. And a flat is slightly lower in pitch. A note of B flat is one half step lower than a B. Music theory is far too complex to go into every detail here, but with time signatures, major scales, minor scales, fifths, harmonics, "raising a third," and "going up a fifth," it is quite heavy on math.

Students who do well in music tend to excel in math. One theory is that music enhances the neural connections that send information between the two hemispheres of the brain. There is also strong anecdotal evidence that playing a musical instrument and singing may prevent or slow down the ravages of Alzheimer's disease or other dementia. Music involves all areas of the brain. There is much research being conducted right now on this connection.

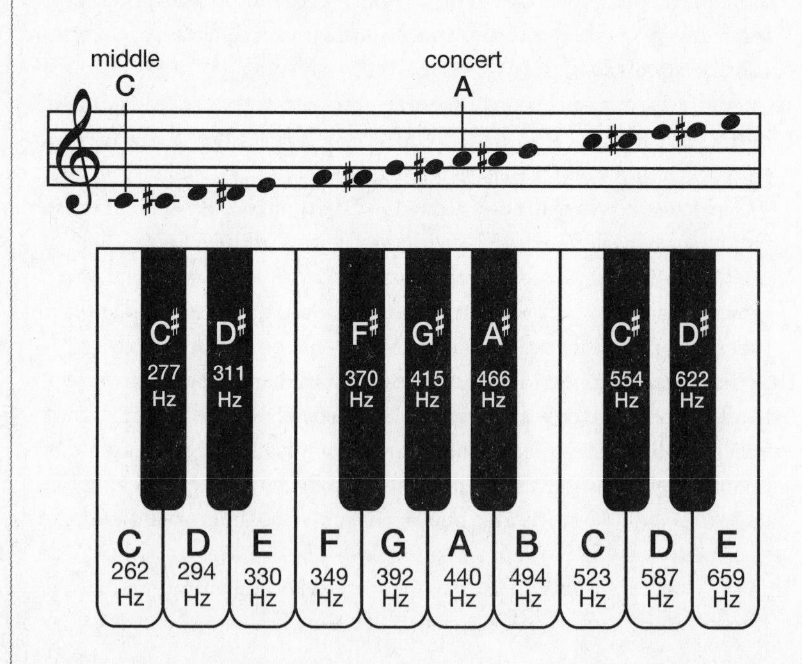

Many famous scientists were musicians. Albert Einstein played the violin, Albert Schweitzer played the organ, and Richard Feynman played bongos and drums.

Last, a joke: What do you get when you drop a piano down a mineshaft? A flat minor.

238. Why does sound travel faster underwater?

The speed of sound through any material depends on the density of the material. In a solid, such as wood or metal, sound waves propagate by vibration. The molecules in a solid are packed tightly together and connected to one another. A passing sound wave vibrates one molecule, and that molecule vibrates its neighbor, and so on, until the energy of vibration dies out.

The same thing happens in a gas, such as air, except the molecules are not attached to each other. The molecules still vibrate, but the travel time is longer, so sound travels slowest in air, faster in water, even faster in wood (11,000 feet per second), and the fastest in metals. The speed of sound in aluminum is about 21,000 feet per second.

The speed of sound in air at freezing (0°C, or 32°F) is about 331 meters per second, or 1,090 feet per second, or about 740 mph.

Now, water is denser than air but less dense than metal. It turns out that sound travels about four times faster in fresh water, slightly more in sea water, than it does in air. That's about fourteen hundred meters per second, or close to 4,400 feet per second. Sound travels about fifteen times faster in steel than it does in air.

The speed of sound in the ocean is quite complex, depending on temperature, pressure, and salinity. While the speed of sound through pure water is 1,400 meters per second, sound travels at 1,522 meters per second through seawater that has 3.5 percent salinity, roughly that of the Pacific Ocean. Sound is also greatly influenced by temperature at the upper regions, where there are sun-warmed layers. The speed of sound in the ocean therefore depends on depth, latitude, and season. At the lower layers of ocean, pressure has a large influence. The higher the pressure, the faster the speed of sound. Such complexities give any submarine captain the fits. That's why expert sonar operators are highly prized in the navy.

Remember those old western movies where the sheriff or marshal would put his ear to the ground and listen for hoof beats? Or to the rail and listen for the far-off train? In addition to traveling faster, sound travels a greater distance and with greater amplitude (loudness) through solid ground or metal rails than through the air.

239. Why is it so quiet after a snowfall?

Ah, a keen observer you are! Yes, it is the epitome of serenity and peacefulness after a fresh snowfall. The newly fallen snow absorbs the sound. A dog's distant bark seems muted and subdued. Passing cars, trucks, and snowplows are barely heard. This snow-induced silence may last for an entire day.

Snow is soft and acts as an effective sound damper. The snow behaves the same way that a carpet does, absorbing more sound and making things quieter than a wood floor.

It is the arrangement of the flakes that damps the sound. Snowflakes come in all shapes and sizes with varying weight and water content. Some crystals are pointy stars, and some are flat plates; each snowflake consists of from one to dozens or hundreds of these crystals.

Snowflakes don't fit together like jigsaw puzzle pieces. They loosely pile up and have plenty of air spaces in between. It is these holes and spaces that absorb the sound, just like the pores in acoustic ceiling tile. Even sound traveling parallel to the snow is absorbed, because the pressure of a passing sound wave pushes air down into the spaces between the flakes.

Later, as gravity compacts the snow and some of the surface crystals melt, it no longer muffles sound effectively. Blowing wind, sunlight, and a tad of rain can all quickly render sound absorption ineffective, too, and you no longer hear that softer, muffled sound. Soon the surface of the snow is more like a whiteboard on the schoolroom wall than like the overhead acoustical tile. The hard-packed snow acts like bare ground, blacktop, or concrete, reflecting and transmitting sound. The noise level returns to normal. John Greenleaf Whittier's 1866 epic poem *Snow-Bound* alluded to the quiet of the countryside after a classic New England snowstorm.

The army uses sound waves to measure snow depth to an accuracy of one inch. The Forest Service, the Bureau of Land Management, the Interior Department, and ski resorts all use this information. Sound waves have increased the accuracy of predictions of spring runoff and flooding. Last, ice thickness in Arctic and Antarctic regions is also monitored in this way.

240. How do compact discs work?

The familiar music CD, and its cousin the DVD, is a remarkable feat of engineering. The compact disc is the most successful electronic product ever introduced. Remember all those assorted ways of listening to music: 45-rpm record, long-playing record (LP), reel-to-reel audiotape, eight-track cassette, and small audiocassette—all with different sizes and speeds?

Some vinyls played at 78 rpm, some at 33⅓, and some at 45 rpm.

In 1982, two giant firms, Sony of Japan and Philips of the Netherlands, introduced the compact audio system. Every disc and every disc player would be made and operated the same and would be compatible worldwide. They finally got it right!

Music is sampled electronically 44,100 times every second. The amplitude of the wave or signal is converted into a number and that number is converted into a binary digital signal, a series of 0s and 1s, by an analog-to-digital converter. These 0s and 1s are recorded on the disc by a series of pits impressed with a master stylus into the plastic surface and arranged in a continuous, long, spiral track from the center of the disc to the outer edge. This surface is then coated with a thin layer of aluminum to reflect a laser beam and overlaid with a lacquer film to protect it.

That laser is in the CD player, whose job is to find and read the data stored on the CD. It focuses the laser on the rows of pits impressed in the disc. The laser beam reflects off the shiny aluminum surface, which is the bottom on the disc, and hits a photocell.

The photocell converts the light into 0s and 1s, the binary signal. The bottom or top of the pit yields a 1, while the edge of the pit is read as a 0. A CD is read from the center of the disc to the outside, just the opposite of its older LP-record cousin. The stream of numbers is read by the laser, changed from light to electricity by a photocell, and thus converted back into music, amplified, and fed to speakers.

The tracks of pits are extremely small. Over six hundred tracks are etched in a width of one millimeter. A millimeter is about the thickness of a dime. The tracks are played back at 1.25 meters, or about four feet, each second, and the rotation speed of the disc decreases from 8.5 revolutions per second when the track is being read near the center of the disc to about 3.5 revolutions per second when read near the outside of the disc. About seventy-four minutes of music or 275,000 pages of text can be stored on a compact disk.

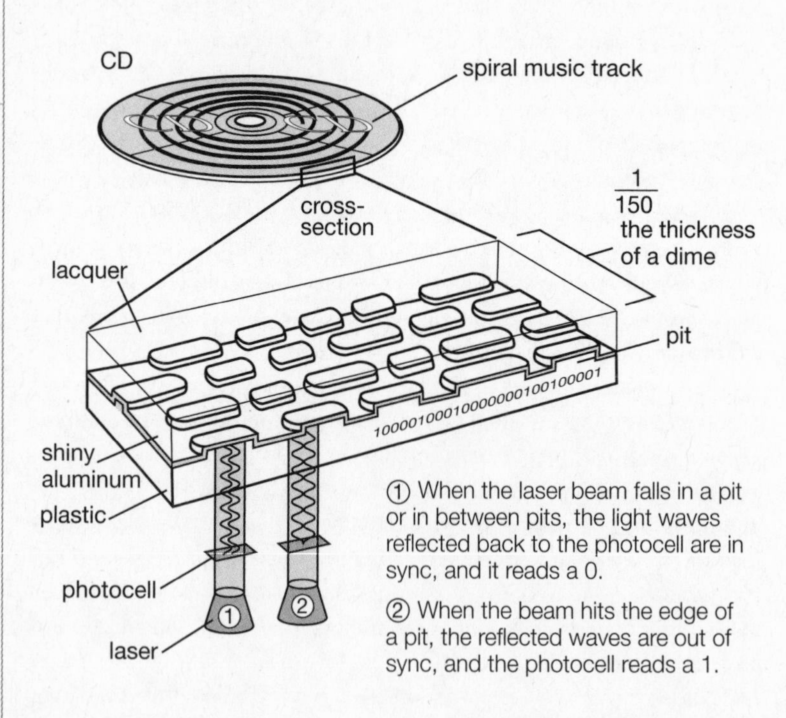

241. What makes sound when two things collide?

Rapid motion of air causes all sounds. When things collide, they vibrate—they flex back and forth. When a piece of material flexes to one side, it pushes the surrounding air particles on that side. These molecules of air collide with the molecules in front of them, pushing them into the next molecules, and so on. This compression of air travels to our ears and pushes our eardrums inward. When the material flexes the other way, it pulls in the surrounding air molecules, creating a partial vacuum. This pressure decrease is called a rarefaction, and when it travels to our ears, it causes our eardrums to move outward. These compressions and rarefactions are what we perceive as sound. Most of the energy of any collision turns into heat energy. Even the bouncing of a ball on the floor causes the ball and floor to heat up slightly. But some of the energy of motion, called kinetic energy, turns into sound energy.

The amount of kinetic energy that goes into sound energy depends on the types of colliding materials and their structural makeup, the speed of the collision. For example, a bullet hitting a tree is much quieter than the same bullet hitting a metal plate. The tree doesn't vibrate very much, but the metal plate does.

The rate at which objects vibrate is known as their frequency, measured in a unit called Hertz or cycles per second. In music we use the term "tone" or "pitch" (see page 35). An object that vibrates at a higher rate has a higher frequency. The amplitude of its vibration determines how loud we perceive the sound to be. The sound level, or amplitude, of sound waves is given in decibels (dB). The ear can hear over a very large range of volume.

The sound of a jet engine is a trillion times as powerful as the least audible sound. A trillion is a one with twelve zeros. Decibels are expressed in logarithmic units, which increase with the exponent of the increase in volume. So a barely audible sound is 0 dB. A sound ten times as powerful is 10 dB. A sound 1,000, or 103, times as powerful as 10 dB is 30 dB. That's how the logarithmic decibel scale works. Normal conversation is 40 to 60 dB. A car horn is over 100 dB. Anything above 85 dB can cause hearing damage over time. Much depends on the duration of exposure. The threshold of pain is 130 dB. At that level most people must cover their ears.

It's about the same as being at a rock concert!

11

At the Fringes of Science

242. Are we alone in the universe?

There have been searches for life on other planets. Two spacecraft named *Viking* landed on Mars in 1976 searching for the building blocks of life. The Viking landers conducted biological experiments. Some of the results were promising but not conclusive. A recent *Phoenix* lander detected perchlorate salts on Mars. But the question of microbial life on Mars remains unresolved. The latest Mars probe, the *Curiosity* rover, may come closer to answering the question about life on other worlds (see page 121).

Decades ago, very few scientists believed there was any life beyond Earth. The prevailing view was that we humans were alone in the universe and life was a chemical quirk that happened only once—an act of God, if you will.

Now the pendulum has swung the other way. The current belief is that the conditions for life are not that hard to duplicate. Life on Earth is based on five elements: carbon, oxygen, hydrogen, nitrogen, and phosphorus. Those elements are known to be plentiful in the universe.

Astronomers have already discovered planets around other suns or stars. NASA recently announced that the *Kepler* spacecraft found several planets similar to the size of Earth orbiting distant stars. Some of these planets are about the right distance from their sun to possibly support life.

Our own Milky Way Galaxy has between two hundred billion and four hundred billion stars. There are an estimated one hundred to two hundred billion galaxies in the known universe. So just based on probabilities, life should be plentiful "out there." But mathematical probabilities do not constitute proof. The Search for Extraterrestrial Intelligence (SETI) is the name of a scientific project being conducted by groups that are listening for radio transmissions from outer space.

Our most likely contact from any intelligent alien life forms will come from listening to their radio transmissions. The book *Contact*, by astronomer and writer Carl Sagan, and the movie by the same name, seems to be the most realistic scenario of what that detection and interaction might look like. Which direction to look in, which radio frequencies to listen on, and which method of communication to use are all unknown. The search has been uncoordinated and sporadic.

243. How does brainwashing work?

Brainwashing is the process of changing the thoughts and beliefs of another person against their will. In psychology, brainwashing is often referred to as "thought reform" or "thought control."

We frequently hear about brainwashing in our everyday lives. Are advertising and infomercials brainwashing? What about political rhetoric? Or rightist talk radio? Most people view these as persuasion, propaganda, education, or campaigning, not as brainwashing in the narrow sense of the word.

Let's look at some of the more famous cases of so-called brainwashing. A few American soldiers captured in the Korean War confessed to their North Korean captors to waging germ warfare, and a few reportedly pledged allegiance to Communism. At least twenty-one refused to return to the United States after 1953. Patty Hearst, heiress to a publishing fortune, was kidnapped by the left-wing Symbionese Liberation Army in 1974. The group isolated and brutalized her, and she ended up joining the group. A famous photo from a surveillance camera shows a gun-toting Hearst robbing a bank.

Lee Boyd Malvo assisted John Allen Muhammad in killing ten people in the 2002 Washington, DC, sniper attacks. Malvo, age seventeen, was abandoned by his mother in Antigua, picked up by Muhammad, brought to the United States, and brainwashed into believing there was an impending war between the Islamic religion and the United States.

Lawyers used the "brainwashing defense" in both the Hearst and Malvo cases. In both trials, the defense claimed their client would not have committed such crimes under normal situations.

Some infamous religious cults could fall into the brainwashing category, too, including the People's Temple with Jim Jones in Guyana; David

Koresh and the Branch Dravidians in Waco, Texas; and Heaven's Gate, founded by Marshall Applewhite, out in San Diego. Some people consider these cases to be brainwashing because the adherents engaged in activities that are out of the mainstream of accepted societal behavior. Examples include isolation, suicide, and total adherence to a single leader.

The Manson family, the Ku Klux Klan, the Unification Church, and the Hare Krishna movement are sometimes put in the brainwashing category, but not everyone would agree.

True brainwashing is an intense form of influence, requiring complete isolation and dependency of the brainwashee. This kind of brainwashing takes place in prison camps or cultist compounds. The practitioner has complete control over the victim's sleep and eating patterns, and even bathroom privileges. Through this forced total dependency, the brainwasher breaks down the person's identity to the point where there's nothing left. The brainwasher then replaces that identity with another set of values, beliefs, and attitudes.

The 1962 psychological-thriller classical movie *The Manchurian Candidate*, starring Frank Sinatra, Lawrence Harvey, Janet Leigh, and Angela Lansbury, is a good depiction of brainwashing. In this political thriller, the son of a prominent US family is brainwashed by his Communist handlers to assassinate a potential political opponent. They remade it in 2004 with Denzel Washington and Meryl Streep. The critics said it was bad, and they are probably right.

244. How does a magician saw a lady in half?

This is a classic magic trick first performed in London in 1920, later executed by Harry Blackstone, Doug Henning, David Copperfield, Criss Angel, and a host of lesser-knowns.

There are several variations on this illusion. The basic trick is one in which the woman lies down in a box. The box, with the woman inside and her head and arms sticking out one end and her feet out the opposite end, is sawed into two halves. Each half is on a dolly or wheeled table. The halves are separated, showing the lady has been sawn into two parts. Then, with great fanfare, the two separated boxes are joined back together. With immense flourish and thunderous applause, the woman rises up out of the box and stands beside the magician.

How is it done? The box is deeper and wider than it appears. The woman climbs in the box, but she folds her legs up into a fetal position, with her head sticking out one end of the rectangular box. Fake legs stick out the other end. So her entire body is actually in one half of the box. She can wiggle the legs with ropes. Some newer illusions even use remote-control radios and motors to operate the legs. And some versions use a second scrunched-up woman on the other side of the box to provide the legs.

The magician, with loud music, much noise, fog, smoke, and wild gestures, "saws" through the box and the woman with a large circular buzz saw or a chainsaw. Metal plates are inserted into grooves, one on each side of the saw cut. The magician pulls the two halves apart and swirls them in circles. The woman smiles, waves her hands, and wiggles her feet. The two halves are joined together, the two metal plates are removed, and the agile young miss pops out of the box, waves again, and bows. The appreciative audience lets out a sigh of relief.

Some magicians actually saw through a piece of wood to enhance the effect and heighten believability. Some station a doctor or nurse nearby. Horace Goldin, in the 1920s, would have an ambulance parked outside or bring one onto the stage if the facilities could handle it. It was Goldin who also made the mistake of taking out a patent on his illusion. In so doing, he revealed how he did the trick. Patents are in the public domain and open to anyone who wants to read them.

Some feminists have criticized the "sawing the lady in half" trick, and a few magicians have responded by placing a male assistant in the box. Magician Dorothy Dietrich uses a male assistant and bills herself as the "First Woman to Saw a Man in Half."

In India, the magician P. C. Sorcar used a buzz saw to cut his wife in two during a televised show. Just as he finished the dastardly deed, the television host quickly signed off and the show ended. The shocked and horrified viewers thought she had accidentally been killed. But the time had simply run out on the live broadcast.

245. If humans could fly, how big would their wings have to be?

Very big! In the 1700s and 1800s, there were people who made wings from wood and fabric, attached them to their arms, and then tried to fly. You see some of those old Movietone News reels from the 1920s and 1930s of guys, wings attached, jumping off bridges and splashing into the water. I say "guys" because women are way too smart to try such stuff!

In order to fly, wings have to create as much lift upward as the person's weight pulls downward. There is a minimum forward speed and a minimum wing area to create the needed lift. The equation for lift is: Lift = ½ v^2A, where = density of the air, v = velocity of the wing or plane, and A = area of the wing. So the bigger the wing (A), the faster the airfoil must move (v). But v, the velocity, is squared, so the power needed for the minimum speed for lift increases more rapidly than the weight of the machine. For example, for a given-size wing, if the weight were doubled, the power would more than double to attain the minimum speed for lift.

When you do the calculations, you find that for a person of average size to fly, their chest muscles would need to project out to about four feet, and their legs would have to be spindly stilts. Not practical at all. Humans were clearly not meant to fly! Remember, we weigh a lot more in proportion to our size than birds, which have hollow bones.

However, human-powered flight is possible. In 1977, Paul MacCready won the one-hundred-thousand-dollar Kremer prize for designing the *Gossamer Condor*, which flew a figure eight, one-mile-long course. MacCready calculated that a good bicyclist could develop, or generate, one-third horsepower almost indefinitely. All he needed to do was find a design that kept the weight down and provided a huge wing, sort of like a super-hang-glider.

Athlete-bicyclist Bryan Allen powered the *Gossamer Condor* on August 23, 1977. He sat in a cockpit of 0.016-inch-thick see-through Mylar, pedaling a bicycle-like mechanism that turned a propeller. The wingspan was ninety-six feet and the plane weighed a mere seventy pounds. The flight took him 7.5 minutes.

Two years later, MacCready designed the carbon fiber *Gossamer Albatross*, which flew the English Channel and won a new $214,000 Kremer prize. The twenty-two-mile distance was covered in a little less than three hours, with a top speed of 18 mph, five feet above the waves. The *Gossamer Albatross* now hangs in the Smithsonian National Air and Space Museum.

246. **Why can't we invent a time machine, and what is a time warp?**

Yes, theoretically, it is possible to travel forward in time, but not backward in time. We can travel faster through time by changing our relative velocity or speed, an idea rooted in Einstein's special theory of relativity (see page 286). To do so would require us to travel as close as possible to the speed of light, which is about 186,000 miles per second. So far, the fastest humans have traveled is seven miles per second, when astronauts were going to and coming back from the Moon. There's a long way to go!

The faster you travel, the slower time moves relative to a stationary observer. Sirius, the Dog Star, the brightest star we see in the night sky, is about nine light-years away. If you traveled at 99.99999999 percent of the speed of light, the relativistic effects are seventy thousand times normal. Mass increases by seventy thousand, time slows down by the same amount, and length contracts by same amount. You could go to Sirius in the morning and get back late at night. You would be less than a day older, but everyone else on Earth would be eighteen years older. There is no magic here, no voodoo of any sort. However, we simply have no way at the present time of getting to those speeds.

A telescope is a kind of time machine. It allows you to look into the past. If you see the Andromeda Galaxy in the night sky, you are looking two million years into the past. The light that you see tonight left Andromeda two million years ago. The Andromeda Galaxy may not even be there now. All we know is how it looked two million years ago.

We can't see or look into the future because it hasn't happened yet, but we can make predictions and conjectures about it. Of course, most of those guesses are predicated on what we already know. And no matter how good we are at predicting it, all of us should care about the future, because we will spend the rest of our lives there!

The term "time warp" originated in science fiction. A time warp is an anomaly, discontinuity, or distortion in the passage of time that would allow events from one time period or era to move to another era. A time warp is an eccentricity in which people and events from one age can be imagined to exist in another age.

Sometimes it refers, sort of jokingly, to the way people live. For example, "Nothing in their lives has changed since the 1950s; they're living in a time warp." Let me tell you, as a teacher who works with teenagers all the time, many teens think their parents are living in a time warp! And they think their grandparents are hopeless!

At times, people are referring to Albert Einstein's concept that time and space form a continuum, which folds, warps, and bends from an observer's point of view. But most of the time the idea of "time warp" refers to the concept that people or a spaceship can travel faster than light travels. Recall those segments in the *Star Trek* television series, where Captain Kirk asked the crew to bring the spacecraft *Enterprise* up to Warp 5. "Warp 5" meant they were traveling 125 times the speed of light according to the old warp table formula.

That's all good in a television program. The reality is quite different. One of the things that Einstein's special theory of relativity shows is that nothing can travel faster than the speed of light.

So just as there is no such thing as teleportation, no "beam me up, Scotty," there is also no way to travel faster than light travels. Light travel is the ultimate speed limit of the universe. If you're going faster than the speed of light, you can forget about the Wisconsin State Patrol. God will pull you over!

Albert Einstein wondered what the world would look like if he could travel on a beam of light. It took him ten years to find out. Answer: He never could, because at the speed of light, time stands still, mass increases to infinity, and length contracts to a narrow line. The warp in space is actually the curvature of space-time, which we commonly refer to as gravity. Isaac Newton explained how gravity behaves, but Albert Einstein explained why gravity is responsible for why any two objects in the universe attract each other.

Time Warp is a science television program on the Discovery Channel featuring an MIT scientist-teacher, Jeff Lieberman, and camera expert Matt Kearney. And "The Time Warp" is a song from the 1975 movie *The Rocky Horror Picture Show*.

247. Is telepathy real?

ESP, or extrasensory perception, is the ability to know something by means other than our five senses. Clairvoyance is the ability to gain information about the location of something or some physical event beyond the normal senses. Psychokinesis, or telekinesis, is the moving of an object by mental power. Mental telepathy, or mind reading, is communication by means other than the five senses. "Paranormal" is an umbrella term for all scientifically inexplicable phenomena, of which there are many. Solid proof of any paranormal powers is very elusive; evidence is largely anecdotal. There seem to be no scientifically accepted cases of mental telepathy, despite many scientific tests.

Many people have had a "telepathic experience" at some point in time. A mother is instantly aware her son has been killed in war. A person has an overwhelming feeling that the phone is about to ring, and it does. Someone thinks about a friend they have not seen in years, and later they run into each other at Walmart. Are these cases of mental telepathy or random coincidences? Scientists don't seem to be able to come up with any repeatable scientific tests to confirm or disavow these phenomena.

There are some people who claim they can move or bend objects using just their thoughts. Uri Geller, an Israeli-British citizen, is famous as a spoon bender, but Geller has been proven to be a fraud. Spoon bending is a common stage trick. Geller was unable to bend any tableware on a Johnny Carson appearance in 1973. Carson, an amateur magician, knew something about trickery.

A lot has to do with what people want to believe. Many people want to believe that there is something out there, that there are powers or forces greater than us. It gives them meaning and purpose for living. For some, it's almost like a religion.

So there will always be folks who think that there is a bigfoot or a

Loch Ness monster, that UFOs landed at Roswell, or that planes and ships mysteriously disappear in the Bermuda Triangle (see page 85). There are people who believe that the landing on the moon was faked. Some still think the Earth is flat.

What about mental telepathy? There may be something to it. But for now, there is not a shred of credible evidence that a person can know what someone else is thinking.

248. How many joules of energy would be fatal to a human?

The joule is a unit of energy. It was named after James Prescott Joule, son of a British brewmaster. Joule measured how much mechanical energy, or work, is required to heat water. He found that 4.2 joules of energy is equal to one calorie of heat energy (see page 238).

So how much is a joule worth? A joule is enough energy to heat up a gram of water 1°C. A joule is the energy needed to lift a small apple up one meter, and it's the energy released when that same apple falls one meter. In electricity, a joule is one watt per second. We pay about nine cents for a kilowatt hour. A kilowatt hour is 3,600,000 joules.

Which gets us around to the question: How many joules can send you to eternity? Some sources state that a discharge of ten joules into the human body can be fatal. Such a source could be from a capacitor used in flash photography or from an electrical wire. Other sources say as little as one joule can be enough to kill a person.

It depends on whether the electricity is delivered to dry or wet skin or under the skin, such as from a cut or open wound. When the skin is wet, it conducts electricity much better (see page 266). If the electricity is delivered across the heart, say from one arm to the other arm, or if it is applied under the skin, very little current is enough to be fatal—as little as twenty milliamperes.

Automated external defibrillators (AEDs) are those portable devices placed in airports, schools, convention centers, and wherever large numbers of people congregate. The idea is that any layman trained in cardiopulmonary resuscitation or first aid can use them quickly. Now they even give instructions to walk you through the process. The first AEDs gave a single, powerful shock of 360 to four hundred joules, but this was too strong and injured some patients. The newer ones are designed to give two sequential lower-energy shocks of 120 to two hundred joules. Each shock moves in the opposite polarity between the pads. The application of the shock from the pads is supposed to stop the arrhythmia, allowing the heart to reestablish an effective rhythm.

Those power strips we buy in stores are now rated in joules. A typical label on an outlet strip might read "Six-Outlet Surge Protector—Rated at 840 Joules." Surge protectors are also rated in joules. The rating tells you how much energy the surge protector can absorb before it fails. A rating of one thousand joules or more is considered good protection. Surge protectors have a component called a metal oxide varistor (MOV) that diverts the extra voltage of a spike or surge to the ground connection.

249. Is spontaneous combustion possible?

Yes, spontaneous combustion does occur. Common examples are rags soaked with gasoline, kerosene, paint thinner, varnish, or car engine cleaners that get tossed into a pile or trash can. The liquids evaporate, and the surrounding air is filled with fumes that are easily combustible. All it takes is some spark or sun beating down of them for a fire to break out.

Another kind of spontaneous combustion is the so-called "grain dust" explosion. These occur around grain elevators, sawmills, and ships being loaded with fertilizer. The air gets filled with microscopic pieces of grain, wood, or powder. Billions of tiny particles provide a huge surface area for burning. Any little spark and kaboom!

But *humans* spontaneously combusting? A lot of people may think so, but scientists aren't convinced. A human can't burst into flame from a chemical reaction within.

But what of the persistent stories, well over a hundred accounts, of finding human remains that appear to have spontaneously combusted? Most, if not all, of these cases were of people who were discovered already dead. In other words, they weren't sitting there in the easy chair watching the Packers game and all of a sudden turned into a human torch. All the cases have a familiar pattern. The victim is almost completely consumed. The coroner discovers a sweet, smoky smell in the room. The extremities often remain intact. The head and torso are burned and charred beyond recognition. The room around the person exhibits little or no sign of any fire.

One of the more common theories is that the fire is caused by methane gas build-up in the intestines. Proteins in the body that are used to speed up chemical reactions (enzymes) ignite the methane gas. Another, more up-to-date theory proposes the wick effect. A candle works by having a burning wick on the inside surrounded by wax made of flammable fatty acids. The wax keeps the candle burning. The wick effect is basically an inside-out candle. In the human body, the body fat supposedly acts as the flammable material, and the victim's clothes act as the wick. As the fat melts from the heat, it soaks into the clothing and acts as a wax-like substance to keep the wick burning slowly. This theory jibes with the fact that the victim's body is destroyed and yet the extremities are hardly touched and the surroundings do not catch on fire.

There is no conclusive proof of spontaneous human combustion. There is, however, much anecdotal evidence. Most scientists say there are likely explanations. Many of the cases have been smokers who fell asleep with a lit cigarette or pipe. A number were under the influence of alcohol, and some had diseases that restricted their movement. Some were fires set by criminals to cover up their dastardly deeds.

Charles Dickens wrote a serialized novel from 1852 to 1853 called *Bleak House*. One of his characters in the book, an alcoholic by the name of Krook, is done away with by spontaneous human combustion.

250. Why do people say our "fate is in the stars"?

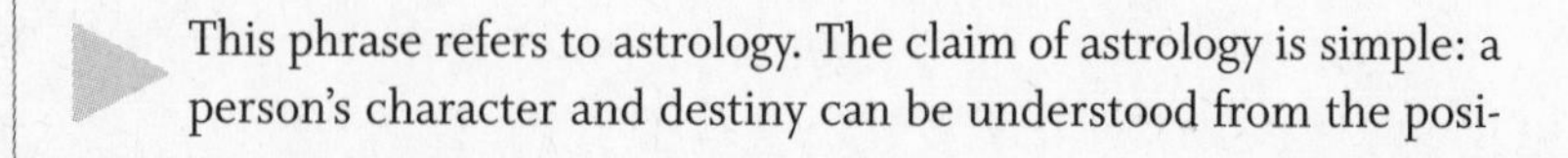

This phrase refers to astrology. The claim of astrology is simple: a person's character and destiny can be understood from the positions of the Sun, Moon, and planets at the moment of their birth. Interpreting the location of these bodies using a chart called the horoscope, astrologers claim to predict and explain the course of people's lives and help them make decisions. Predictions and advice like these for the coming month are also commonly called horoscopes. Astrology arose at a time when humankind's view of the world was dominated by magic and superstition, when the need to grasp the patterns of nature was often of life-and-death importance. Astrologers believe that the important constellations are the ones the Sun passes through during the course of a year. These are the constellations of the zodiac.

Simply put: astrology doesn't work. Many careful tests have shown that, despite their claims, astrologers really can't predict anything. French statistician Michel Gauquelin sent the horoscope of one of the worst mass murderers in French history to 150 people and asked how well it fit them. Ninety-four percent of the subjects said they recognized themselves in the description. Researcher Geoffrey Dean reversed the astrological readings of twenty-two subjects, substituting phrases that were the opposite of what the horoscopes actually stated. Yet the subjects in this study said the readings applied to them just as often (95 percent of the time) as people for whom the original phrases were intended. No wonder astrological predictions are written in the most vague and general language possible.

We see a lot of TV psychics and tarot card readers willing to part us from our money. I don't think we should be tied to an ancient fantasy, left over from a time when humans huddled by the campfire, afraid of the night. Have fun by reading your horoscope in the daily newspaper, but don't take any stock in it. Have you noticed that the horoscope is often placed on the same page as the comics? Remember that line from Shakespeare's *Julius Caesar*, spoken by Cassius: "Men at some time are masters of their fates. The fault, dear Brutus, is not in our stars, but in ourselves."

Acknowledgments

I would like to thank the many individuals who helped make this book possible. My wife, Ann, has been very supportive and encouraging. The many long hours of research and writing can cut into the hours of family time.

Many thanks to the several thousand students of all ages, as well as adults, who sent questions to me. Many of these questions came via their teachers. A special thank-you to you teachers, especially Rock Shutter, for sharing the contents of your students' querying minds.

Thanks to the many fellow teachers and friends who critiqued some of my answers to questions. The same thanks goes to my three brothers and five sisters. I have turned to several medical personnel for advice, namely Dr. Scott Nicol, Dr. Alan Conway, Dr. Rod Erickson, and Dr. Rick Erdman. I received excellent advice from engineering instructors at UW–Milwaukee.

A big thank-you to the four thousand students that were enrolled in my science classes at Tomah High School, Tomah, Wisconsin. It was an honor to have you in my physics class for one or two years. I have many fond memories from close to forty years of teaching.

Thanks to Amy Fass, whose many hours of research and fact-checking improved the accuracy of this book. Thanks also to Karen Giangreco and Ruth Murray for providing excellent sketches and illustrations.

Finally, I am grateful to Matthew Lore, president and publisher of The Experiment, and his team, especially my editor, Nicholas Cizek, who provided suggestions for improvements and encouraged changes where needed. *Ask a Science Teacher* would not have been possible without their guidance.

About the Author

Larry Scheckel taught high-school-level physics and aerospace science for more than thirty-eight years. A three-time Tomah (Wisconsin) Teacher of the Year and six-time Presidential Awardee at the state level, Scheckel has shared science with thousands of adults and students in presentations at such venues as Boys & Girls Clubs of America, Rotary Clubs, children's museums, and conventions. He lives with his wife, Ann, in Tomah.